ANCIENT WISDOM AND MODERN MISCONCEPTIONS

A Critique of Contemporary Scientism

BY THE SAME AUTHOR

Cosmos and Transcendence:
Breaking Through the Barrier of Scientistic Belief

Science and Myth:
With a Response to Stephen Hawking's The Grand Design

The Quantum Enigma: Finding the Hidden Key

Sagesse de la Cosmologie Ancienne

Christian Gnosis:
From Saint Paul to Meister Eckhart

Réponse à Stephen Hawking:
De la Physique à la Science-Fiction

Theistic Evolution:
The Teilhardian Heresy

Rediscovering the Integral Cosmos:
Physics, Metaphysics, and Vertical Causality
(with Jean Borella)

Physics and Vertical Causation:
The End of Quantum Reality

Wolfgang Smith

ANCIENT WISDOM
AND MODERN MISCONCEPTIONS

A Critique of Contemporary Scientism

ANGELICO PRESS
SOPHIA PERENNIS

First published in the USA
© Wolfgang Smith 2013
Expanded edition © Wolfgang Smith 2015
Angelico Press / Sophia Perennis, 2015
Revised and expanded edition of the work originally
published as *The Wisdom of Ancient Cosmology:
Contemporary Science in the Light of Tradition*
Foundation for Traditional Studies, Oakton, VA, 2004

Series editor, James R. Wetmore

For information, address:
Angelico Press / Sophia Perennis
4709 Briar Knoll Dr.
Kettering, OH 45429
www.angelicopress.com

978-1-62138-021-4 pb
978-1-62138-023-8 cloth

Cover Design: Michael Schrauzer
Cover image: Thomas Wright, "A synopsis of the universe, or,
the visible world epitomiz'd" (details), [London]: Printed for
the Author and According to Act of Parliament, 1742.

TABLE OF CONTENTS

In Memoriam
Rev. Fr. Malachi Martin
† July 27, 1999

Preface

This is a revised edition of a book entitled *The Wisdom of Ancient Cosmology* published a decade ago, containing eleven of the original twelve chapters. These take the form of essays covering a broad spectrum of topics—from the seeming paradoxes of quantum theory to the staggering claims of contemporary astrophysics—each of which hinges upon an application of "Ancient Wisdom" to the topic at hand: "the rigorous application of traditional keys" to put it in Professor Borella's words.

The customary order of priority is thus reversed: instead of viewing the sapiential traditions from the perspective of contemporary science—which amounts to "explaining away" the most profound and most sacred beliefs of mankind!—the new approach mobilizes the aforesaid Wisdom to attain a *metaphysical* grasp of the stipulated scientific findings. The result is twofold. First, this approach brings to light the crucial distinction between actual scientific discovery and what I have termed "scientistic myth," which turns out to include a major part of what the general public regards as scientific fact. And secondly, having separated the grain from the chaff one arrives at the realization that the actual findings of contemporary science—its "non-mythical" component thus—so far from contradicting the traditional doctrines, can indeed be integrated into that *cosmologia perennis*, which is to say: can in fact be *understood*.

Here, then, we have the answer to Feynman's "No one understands quantum theory" and Whitehead's lament that physics has turned into "a kind of mystic chant over an unintelligible universe." What renders the universe intelligible proves finally to be none other than that long-neglected and long-despised Ancient Wisdom which in the final count outweighs the contingent productions of mortal endeavor.

WOLFGANG SMITH
Camarillo, December 15, 2012

1

Foreword

That there are today, in our civilization, religions with followers still standing by their beliefs is, with respect to the modern world, a kind of anomaly: religious belief definitely belongs to a bygone age. A believer's situation, whatever his religion, is not an easy one then. But what is true for all sacred forms is especially true for Christianity, because for three centuries it has been directly confronted by the negations of modernity. The day when Hinduism, Buddhism or Islam experience the omnipresence of this modernity, they will undoubtedly in their turn undergo serious crises.

The blows dealt by the modern world against a people's religious soul is in the first place concerned with the plane of immediate and daily existence. No need for ideological struggle here; merely by the strength of its presence and extraordinary material success, this world refutes the world of religion, silences it, and destroys its power. This is because religion speaks of an invisible world, while contemporary civilization renders the sensory world more and more present, the invisible more and more absent.

This is, however, only the most apparent aspect of things. The omnipresence of a world ever more "worldly" is only the effect, in the practical order, of a more decisive cause that is theoretical in nature, namely the revolution of Galilean science, its technical progress being only its consequent confirmation. For the religious soul, the importance of the scientific revolution consists in the fact that it affects this soul's own inwardness. As powerful as it might be, for the human being, society represents only an environment which it can in principle ward off. Whereas the scientific revolution, insofar as it ascribes the truth to itself, imposes itself irresistibly and from within on the intelligence that it besieges. It is a cultural and therefore a "spiritual" revolution to the extent that it makes an appeal to our mind. But whenever it is a question of a believer's mind, it is the vision of the world and the reality implied by his faith that is subverted. What remains then is the option either to renounce his faith, or else—an almost desperate solution—to renounce entirely the cosmology that it entails.

On the whole Christian thought has committed itself to this second way: to keep the faith (but a "purified" faith!) and abandon all the cosmological representations by which that faith has been expressed. This is a desperate solution because these cosmological representations are first scriptural *presentations*, the very forms by which God speaks to us about Himself. But if we disregard these forms, what remains of our faith? Scripture informs us that the apostles saw Christ raised from the earth and disappear behind a cloud, while Galilean science objects that space is infinite, that it has neither high nor low, and that this ascension, even supposing it to be possible—which it is not—is meaningless. What remains is then to see in it a symbolic fiction by which the early Christian community attempted to speak its faith in a vanished Jesus Christ: if He is no longer visible, this is because He has "gone back to heaven." Following Rudolf Bultmann the majority of Protestant and Catholic exegetes and theologians have adopted this "solution." Since then an immense process of *demythologization* of Christian scriptures has been in progress. According to Bultmann, what is mythological is a belief in the objective reality of revelation's cosmological presentation: "descent," "resurrection," "ascension," etc. To demythologize is to understand that this cosmological presentation is, in reality, only a *symbolic* language, in other words, a fiction. To pass from myth to symbol, this is the hermeneutic that enables a modern believer, living at the same time in two incompatible universes—that of the Bible and of Galilean science—to avoid cultural schizophrenia.

But at what price? At the price of making unreal all biblical teachings on which faith relies and with which it is bound up. To reject this cosmological presentation, the witnesses of which the apostles, for example, vouch to have been, is this not to reject with the selfsame stroke the faith attached to it? What does this parting of faith from its cosmological garment, of kerygma from myth, imply? Basically, would this not separate the Divine Word from its carnal covering and ultimately deny the Incarnation?

How surprising that another way never occurred to Bultmann, a way which, had it been taken into consideration, might have changed many things in the course of the West's religious history. It is this way that the distinguished mathematician Wolfgang Smith proposes to explore, and into which he now offers us insights. In the present crisis, in which Christian thought is split between an impossible fideism and its confinement to moral problems, his book discloses a liberating perspective which, in the name of science itself, restores to faith its entire truth. It would be hard to exaggerate the importance of such a work. On the most essential points, the most burning questions concerned with biblical cosmology, heliocentrism, the nature of space and matter, the concept of a true causality, etc., Wolfgang Smith shows how the conclusions of contemporary science cease to be incompatible with the affirmations of traditional cosmology.

I will only mention the admirable analyses developed in Chapter 6, "The Pitfall of Astrophysical Cosmology." First he sets forth the criticisms certain scientists have directed at the major dogma of the new cosmology, which is "big bang" theory: these criticisms reveal its weakness and even impossibility, and thus disqualify the use theologians have made of that theory. Next he establishes that physics, when applied to celestial bodies, having *de facto* no operational value in that domain, has necessarily an ontological significance, which however is illegitimate. If in fact sidereal bodies, as required by quantum theory, are composed of an almost nonexistent dust of particles, these bodies themselves, as identifiable realities, vanish into space. For a *body* (un *corps*) is also *a* body (*un* corps); a being that is not *a* being is no longer a *being*, says Leibniz. Now quantum theory has nothing to say about the existence of this unitary principle needed to account for the reality of a body: it is therefore truly incapable of accounting for the reality of any corporeal being, be it stellar or earthly (which is why some physicists have fallen into an idealism insupportable in other respects). Hence it is absolutely necessary, as Wolfgang Smith reminds us, to have recourse to what traditional philosophy calls a "substantial form," a unitary principle that endows a material body with its own reality. This is no speculative luxury that might be dispensed with, but a rigorously scientific need, since it is the incontestable truth of quantum physics itself that, for want of this substantial form, the reality of bodies is rendered forever *inexplicable* and indeed *impossible.*

We should be thankful to Professor Wolfgang Smith for having reminded us of these primary truths with the authority of a recognized scientist and the full resources of his broad philosophic and religious culture. I also salute his courage, for he has dared to confront, with such constancy, the dominant ideology of modern culture, which is not without risk, to say the least. This ideology has turned science (a certain kind of science!) into the official mythology of our times. Basically, Wolfgang Smith shows us, with simplicity and sometimes with much humor, that Bultmann has chosen the wrong object: it is not religion but the customary interpretation of science that needs to be "demythologized." Now, only the doctrine of the *philosophia perennis* is able to accomplish this, and thereby to disclose the full truth of science itself. And to this end I do not think there is a more useful and efficacious work than the one by Wolfgang Smith that I have the pleasure of prefacing.

JEAN BORELLA
Université de Nancy

Introduction

Given that the tenets of "Ancient Wisdom" will prove pivotal, it behooves us to clarify that concept right from the start. The point needs first of all to be made that the "wisdom" we take to be sovereign is so not simply by virtue of being ancient, but that its primacy derives rather from the fact that it is *traditional*. Yet it happens that the ancient cosmologies did tend to be traditional in varying degrees, and that we shall consequently be concerned with "ancient wisdom" after all.

What renders a doctrine traditional is however the fact that it is *more* than "historical": that in truth it derives from a *revelation*, of which it constitutes the embodiment. Yet what exactly this means remains perforce incomprehensible so long as we position ourselves within the confines of the prevailing worldview. Suffice it to say that authentically traditional doctrine transgresses the plane of mundane discourse, that the very bounds of space and time which define *our* world are somehow breached. Even when it refers to ordinary things, it does so from an extraordinary perspective: *sub specie aeternitatis* as the Scholastics say. It possesses, moreover, power to inspire: to guide step by step from "*the things that are made*" to the "*invisible things of God*"[1] from which in fact they ultimately derive. Such a teaching serves thus to reconnect its votaries to the spiritual order, and is therefore "religious" in precisely the original sense of *re-ligare*, of "binding back."

Ancient cosmologies tended to be traditional in varying degrees, we have said; yet it may be more accurate to describe them as traditional doctrines in process of decay. There has been a progressive "falling off" from the heights of revelation, due evidently to the collective human propensity to be forgetful of things spiritual, uncomprehending of the higher significations concealed, as it were, within

---

1. Rom. 1:20.

7

teachings of an authentically traditional kind. Due ultimately to what St. Paul terms a *"darkening of the heart,"* the spiritual content of sacred doctrine becomes thus progressively obscured. And this appears to be true above all in the case of traditional cosmology, the spiritual dimensions of which have become almost totally forgotten since the so-called Enlightenment. We need however to bear in mind that this collective process of obscuration did not begin with Galileo, Newton or Descartes, but traces back to the earliest historical periods, and was already well in progress even while the cosmologies in question were still accepted as normative in their external sense. One knows, however, that "the letter killeth," and that the outer sense of a sacred doctrine cannot outlive for too long the neglect of its inner dimension. It is perhaps surprising, therefore, that "archaic" cosmology did survive in Europe, at least in some of its outer forms, for as long as it has: roughly till the eighteenth century, when it came to be replaced by paradigms of a very different kind.

In light of these observations it is apparent that one cannot expound or delineate traditional cosmology as one would, say, the facts of botany or the history of Greece. Yet there are principles to which all traditional cosmology conforms, and these *can* be stated, and can serve from the start as guideposts along the path of discovery. I propose now to enunciate three such principles, in a kind of ascending order.

The first is quite simple: it affirms that traditional cosmology has to do primarily with the *qualitative* aspects of cosmic reality: the very components, thus, which modern cosmology excludes categorically. As we shall have occasion to see before long, this first recognition, simple though it be, has already enormous implications.

The second principle relates to the metaphysical notion of "verticality" and affirms a hierarchic order, in which the corporeal domain, as commonly understood, constitutes in fact the lowest tier. The transition from traditional to contemporary cosmology entails thus a drastic diminution: an ontological shrinkage of incalculable proportions, which of course pertains, not to the cosmos as such, but to the horizon of our worldview. As if to compensate for this reduction, contemporary cosmology imputes spatio-temporal magnitudes to the universe at large that stagger the imagination by their sheer quantitative immensity. The fact remains, however, that the spatio-temporal universe in its entirety constitutes but the outer shell, so to speak, of the integral cosmos as conceived in traditional cosmology.

The third principle presupposes the preceding two, and affirms that man constitutes a microcosm or "universe in miniature," which in a way recapitulates the order of the integral cosmos itself. This recognition, moreover, might well be singled out as the defining characteristic of the traditional worldview, which can in truth be characterized as "anthropomorphic." I say "in truth," because what stands at issue is an ascription of anthropomorphism which is not merely poetical or imaginary, but factual. Tradition maintains that man and cosmos exemplify, so to speak, the same blueprint, the same master plan. This means, first of all, that even as man is trichotomous, consisting of *corpus, anima* and *spiritus,* so too does the cosmos prove to be tripartite, consisting of what Vedic tradition terms the *tribhuvana,* the "three worlds." Generally speaking, all that is essential in the cosmos at large has its counterpart in the *anthropos.*

It may be noted that whereas this conception is common to the major sapiential traditions of the world, it "comes into its own" when interpreted in a Christian key. It is *in Christ,* after all, the Incarnate Word, that we encounter the *anthropos* in his untruncated fullness; and did not St. Paul declare that "*in Him resides all the fullness of the Godhead bodily*"?[2] Yes, "*bodily*" (*somatikos*)! Now this in itself implies that there can be nothing *essential* in the cosmos which is not to be found in the Body of Christ. Man is by no means a stranger in a hostile or indifferent universe, but constitutes *de jure* the very heart and center of the cosmos in its entirety.

Yet it hardly needs saying that this doctrine flies in the face of all contemporary belief: that according to our scientific wisdom, so far from constituting a microcosm, man represents actually a most unlikely anomaly, a precarious molecular formation of astronomical improbability. Except for the laws of physics and chemistry, which

2. Col. 2:9.

are presumably operative in the cells of his body even as they are in stars and plasmas, he enjoys no kinship whatever with the universe at large, which presents itself as indifferent and ultimately hostile to his human aspirations.

It is no wonder then, in light of what has been noted above, that this presumed scientific wisdom runs aground when it comes to the understanding of man himself: that it has in fact shown itself categorically incapable of accounting for even the most rudimentary act of cognitive sense perception, let alone for the higher modes of sensory and intellective knowing.[3] I would like from the start to call attention to the fact that the very possibility of human knowing demands a certain kinship between man and cosmos: the very kinship, in fact, affirmed by the traditional doctrine. As Goethe has beautifully put it: "*Wäre das Auge nicht sonnenhaft*" ("If the eye were not sunlike") we could not behold the Sun. Strange and indeed unbelievable as it may sound: man is able to know the cosmos because, in a marvelous way, the cosmos pre-exists in him.[4]

Having broached the subject of traditional cosmology, we need of course to ask ourselves whether that cosmology is compatible with the findings of contemporary science: its actual findings, namely, as distinguished from the surrounding nimbus of scientistic beliefs. Given what we do know today about the universe—its origin, its configuration, and its laws—is it logically defensible to maintain the principles and tenets of that traditional cosmology? To be sure, an overwhelming majority of scientists, philosophers and theologians would instantly answer in the negative. They take it to be self-

---

3. I have discussed this question at length in Chapters 4 & 5 of *Science and Myth* (Tacoma, WA: Angelico Press/Sophia Perennis, 2012).

4. It should be pointed out that an isomorphism of uncanny accuracy between macro- and microcosm has been uncovered some decades ago by a German phenomenologist named Oskar Marcel Hinze, which, perhaps for the first time, supplies hard evidence *of a scientific kind* in support of the traditional contention. On this subject I refer to *Science and Myth*, op. cit., Chapter 6.

evident that modern science has, once and for all, disqualified the "primitive conjectures" of pre-modern man. This judgment accords, moreover, with the evolutionist outlook which perceives everything as arising "from below," and is consequently disposed to give pride of place to the latest turn of collective human belief.

Other groups, comprising a less numerous category, profess high respect for ancient doctrine while they implicitly deny its truth. I am thinking especially of individuals bent upon "psychologizing" every facet of ancient cosmological belief, as if it were a question simply of human fantasies. Suffice it to say that nothing could be more radically opposed to the traditional cosmological teachings, which invariably uphold the basic distinction between the human and the cosmic realms, and insist that cosmology refers indeed to the second of these domains; and one might add that the traditional conception of man as microcosm does not alter this fact, but presupposes it rather. This brings us finally to a third group, which seems to take the ancient doctrines at their word while likewise accepting the outlook of contemporary science, as if there were not the slightest conflict or incompatibility between their respective claims. I have in mind, for example, individuals who cheerfully cast horoscopes and interpret these in more or less traditional terms without realizing that this makes little sense in an Einsteinian universe.

Diverse as these respective mentalities may be, they exhibit a common deficiency. What I find conspicuously lacking is the slightest mark of critical acumen, the least sign that a critique of the prevailing *Weltanschauung* has taken place. Yet it should be abundantly clear that a critique which penetrates to the very foundations of that worldview constitutes today the *sine qua non* for a sane approach to cosmology. Whatever one may think about the past, we live in a present dominated intellectually by the science of our day, and that science needs to be deeply probed and in a way transcended before one can access whatever traditional wisdom the past may hold. As Theodore Roszak has sagaciously observed: "Science is our religion, because we cannot, most of us, with any living conviction, *see around it.*" That is the overriding fact; and it hardly matters whether we extol the wisdom of the past and cast horoscopes: so long as we do not "see around" that science—"break through the barrier of scientistic belief" as the subtitle of my first book has it—we remain intellectually modern, profane, and in truth anti-traditional. And let me add parenthetically that this may be the reason why there are today few if any saints in the cadre of St. Augustine or St. Thomas Aquinas: it seems to be almost a precondition for sanctity, these days, to have escaped a university education.

My point, in any case, is that "the barrier of scientistic belief" can indeed be breached, which is to say that it *is* possible to overcome the preconceptions of modernity and postmodernity alike, and in so doing rejoin the pre-modern human family. This does not of course confer instant illumination; what it does bestow, however, is a faint vision, a fleeting glimpse at least, of high and sacred truths; and this is something priceless and irreplaceable, and greater by far than any imagined wisdom tinged with contemporary preconceptions.

Implicit in all I have said is the contention that the principles and tenets of traditional cosmology have not in fact been disqualified by the actual findings of contemporary science; and to verify this claim one needs to engage in the kind of critique, alluded to above, whereby one arrives at a sharp separation of scientific fact from scientistic fiction. Yet this is only half of what needs to be done: for it is likewise imperative to interpret what science *has* disclosed, to make sense out of its positive findings, failing which one inevitably succumbs once more to some kind of scientistic fantasy. I contend, however, that to arrive at an authentic interpretation of contemporary science one requires the resources of traditional doctrine itself. In an earlier monograph, entitled *The Quantum Enigma*, I have carried out an approach of this kind for physical science as such, with the result that its generic object—namely the physical universe, properly so called—could be integrated into the traditional ontologies as a sub-corporeal domain. And as might be expected, this throws light on many questions and explains findings which hitherto had actually seemed paradoxical. As a rule one finds that the very discoveries of physics which strike us as the most bizarre are those that harbor a major metaphysical truth. Such is the case above all when it comes to the quandaries of quantum theory, the uncanny behavior of quantum particles: what renders the theory paradoxical, it turns out, is a failure on the part of the physicist to distinguish between two disparate ontological planes: the *corporeal*, namely, and the *physical*.[5] Nothing that is true, nothing that is valid, is lost by adopting a traditional outlook: from a higher point of vantage one sees, not less, but more. *The teachings of the traditional schools, I say, so far from being disqualified by the discoveries of contemporary science, are in fact needed to arrive at a proper understanding of science itself.*

5. See Chapter 3 of *The Quantum Enigma* (Tacoma, WA: Angelico Press/Sophia Perennis, 2012), first published in 1995. Since that time I have explained this fundamental point repeatedly (for instance, in Chapters 2, 3, and 7 of *Science and Myth*, op. cit.). We will be dealing with this question once more in the first Chapter of this book.

1

From Schrödinger's Cat
to Thomistic Ontology

Templeton Lecture on Christianity and the Natural Sciences
Gonzaga University, 1998

Let me call your attention, first of all, to an as yet largely unobserved fact: while the scientific worldview continues to consolidate its grip upon society, something quite unexpected has come to pass. The decisive event occurred almost a century ago in fact, back in the early decades of the twentieth century. Since then that so-called scientific worldview—which to this day reigns as the official dogma of science—no longer squares with the known scientific facts. What has happened is that discoveries at the frontiers of science do not accord with the prevailing *Weltanschauung*, with the result that these findings present the appearance of paradox. It seems that on its most fundamental level, physics itself has disavowed the very worldview proclaimed in its name. This science, therefore, can no longer be interpreted in the customary ontological terms; and thus, as one quantum theorist has put it, physicists have "lost their grip on reality."[1] But obviously this fact has not been publicized, and as the aforesaid physicist observes, constitutes indeed "one of the best-kept secrets of science." It needs however to be pointed out that, strictly speaking, physics did not "lose" its "grip on reality": in light of the new findings the fact is rather that modern physics never *had* such a "grip" in the first place. This Baconian science, rigorously conceived—that is to say, interpreted without recourse to the cus-

1. Nick Herbert, *Quantum Reality* (Garden City: Doubleday, 1985), p.15.

14

tomary penumbra of scientistic beliefs—reduces quite simply to a positivistic discipline. And this explains Whitehead's famous description of that science as "a kind of mystic chant over an unintelligible universe,"[2] as well as the admission by one of the leading quantum theorists that "no one understands quantum mechanics." To be sure, the incomprehension to which Feynman alludes refers to a *philosophic* plane: one understands the mathematics of quantum mechanics and its connection with empirical procedures, but not the ontology.

Broadly speaking, physicists have reacted to this impasse in three principal ways. The majority, perhaps, have found comfort in a basically pragmatic or "operational" outlook—the fact that "it works"— while some persist, to this day, in the patently futile attempt to fit the positive findings of quantum mechanics into the pre-quantum scientistic ontology. The third category, lastly, which includes some of the most eminent names in physics, convinced that the pre-quantum ontology is now defunct, have cast about for new philosophic postulates, in the hope of arriving at an acceptable conception of physical reality. There appear to be a dozen or so worldviews presently competing for acceptance in the upper reaches of the scientific community, which to the uninitiated seem to range quite literally from the bizarre to the outright ridiculous.

It is not my objective, in this lecture, to regale you with yet another *ad hoc* philosophy designed to resolve or explain away quantum paradox. I intend rather to do the very opposite: to show, namely, that there is absolutely no need for a new philosophic *Ansatz*, that the problem at hand can in fact be resolved quite naturally on strictly *traditional* philosophic ground. What I propose to show is that the quantum facts, divested of scientistic encrustations, fit perfectly into a very ancient and venerable ontology: the Thomistic, namely, which as you know, traces back to Aristotle. Rejected by Galileo and Descartes, and subsequently marginalized, this reputedly outmoded medieval speculation, it turns out, resolves the issue instantly. No need for *ad hoc* postulates that stagger our understanding: the keys for which physicists have been

2. *Nature and Life* (New York: Greenwood Press, 1968), p.15.

groping since the advent of quantum theory, it turns out, have been at hand for well over two thousand years.

First formulated in 1925, quantum mechanics has shaken the foundations of science. It appears as though physics has, at long last, broken through to its own fundamental level; it has discovered what I shall henceforth term the *physical universe*—a realm which seems to defy some of our most basic conceptions of objective reality. It is a world (if we may call it such) that can be neither perceived nor imagined, but only described in abstract mathematical terms. The most useful and widely accepted of these representations is the one formalized in 1932 by the Hungarian mathematician John von Neumann. In this model the state of a physical system is represented by a vector in a so-called complex Hilbert space. This means, basically, that a state can be multiplied by a complex number, and that two states can be added, and that non-zero linear combinations of states, thus formed, will again be states of the physical system. Now, it is this fundamental fact, known as the superposition principle, that gives rise to quantum strangeness. Consider, for instance, a physical system consisting of a single particle, and then consider two states, in which the particle is situated, respectively, in two disjoint regions A and B, which can be as widely separated as we like. A linear combination of these two states with non-zero coefficients will then determine a third state, in which apparently the particle is situated, neither in A nor in B, but somehow in both regions at once. Now, one may say: "State vectors actually describe, not the physical system as such, but our knowledge concerning the physical system. The third state vector, thus, simply signifies that, so far as we know, the particle can be in A or in B, with a certain probability attached to each of the two possible events." A grave difficulty, however, remains; for the state of the physical system corresponding to the third state vector can in fact be produced experimentally, and when one does produce that state one obtains interference effects which could not be there if the particle were situated in A or in B. In some unimaginable way the particle seems thus to be actually in A and B at once.

What happens, then, if one measures or observes the position of the particle in the third state? It turns out that the act of measurement instantly throws the system into a new state. The *detected* particle, of course, is situated either in A or in B, which is to say that only unobserved particles can bilocate. All this, to be sure, is very strange; but let me emphasize that from a mathematical point of view all is well, and that in fact the theory functions magnificently. As I have said before, what puzzles physicists is not the mathematics, but the ontology.

Thus far I may have conveyed the impression that superposition states are rare and somehow exceptional. What is indeed exceptional, however, are states in which a given observable *does* have a precise value (the so-called eigenstates); yet even in that case it happens that the system remains necessarily in a superposition state with respect to other observables. The quantum system, thus, is always in a state of superposition; or more precisely, it is at one and the same time in many different states of superposition, depending upon the observable one has in view. On the quantum level superposition is not the exception, but indeed the fundamental fact.

At this point one might say: "There is no reason to be unduly perplexed; superposition applies, after all, to microsystems too minute to be perceptible. Why worry then if 'weird things' happen on the level of fundamental particles and atoms? Why expect that one can picture things or happenings which are by nature imperceptible?" Most physicists, I believe, would be happy to adopt this position, if it were not for the fact that superposition tends to bleed into the macroscopic domain. It is this quantum-mechanical fact that has been dramatized by Erwin Schrödinger in the celebrated thought-experiment in which the disintegration of a radioactive nucleus triggers the execution of the now famous "Schrödinger cat." According to quantum theory, the unobserved nucleus is in a superposition state, which is to say that its state vector is a linear combination of state vectors corresponding to the disintegrated and undisintegrated states. This superposition, moreover, is transmitted, by virtue of the experimental setup, to the cat, which is consequently in a corresponding superposition state. In plain terms, the cat is both dead and alive! The hapless creature remains moreover in this curious condition until an act of observation "collapses its state vector" as the expression goes, and thereby reduces that state vector to one or the other classical eigenstate. The cat is then either dead or alive, period.

The mystery here, of course, has nothing to do with the nature of cats, but pertains rather to the role of measurement in the economy of quantum mechanics. Now, measurement is a procedure in which a given physical system is made to interact with an instrument so that the resultant state of that instrument indicates the value of some specific observable associated with that system. For example, a particle is made to collide with a detector (a photographic plate, perhaps) which registers its position at the moment of impact. Prior to this interaction, the particle will in general be in a superposition state involving multiple positions; we must think of it as spread out over some region of space. Its evolution or movement prior to impact, moreover, is governed by the so-called Schrödinger equation, which is linear, and hence preserves superposition, and is moreover strictly deterministic, which means that given an initial state, the future states are then uniquely determined. At the moment of impact, however, this deterministic Schrödinger evolution is superseded by another quantum-mechanical law, a so-called projection, which singles out one of the positions represented in the given superposition state—apparently for no good reason!—and instantly assigns the particle to the chosen location. Now, this simple scenario exemplifies what happens generally in the act of measurement: a physical system interacts with an instrument or measuring apparatus, and this interaction causes the Schrödinger evolution of the system to be superseded by an apparently random projection. It is as though the trajectory of a particle, let us say, were suddenly altered without any assignable cause. Why does this happen? Inasmuch as the instrument is itself a physical system, one would expect that the combined system, obtained by including the instrument, should itself evolve in accordance with the corresponding Schrödinger equation; but in fact it does not! What is it, then, that distinguishes the kind of interaction we term measurement from other interactions between physical systems, in which Schrödinger evolution is *not* superseded?

Quantum theory holds many puzzles of this kind; the "scandal" of superposition assumes many forms. I would like to mention one more of these enigmas, which strikes me as particularly significant. One might think of it as a simplified version of the Schrödinger cat paradox. In the words of Roger Penrose, the problem is this: "The rules are that *any* two states whatever, irrespective of how different from one another they may be, can coexist in any complex linear superposition. Indeed, any physical object, itself made out of individual particles, ought to be able to exist in such superpositions of spatially widely separated states, and so be 'in two places at once'! . . . Why, then, do we not experience macroscopic bodies, say cricket balls, or even people, having two completely different locations at once? This is a profound question, and present-day quantum theory does not really provide us with a satisfying answer."[3] It happens that these matters have been debated for a very long time, and various interpretations of the mathematical formalism have been proposed in an effort to make philosophic sense out of the theory. However, as Penrose observes: "These puzzles, in one guise or another, persist in *any* interpretation of quantum mechanics as the theory exists today."[4] After more than half a century of debate it appears that no clear resolution of the problem is yet in sight. One thing, however, one crucial point, has been consistently overlooked; and that is what I must now explain.

As one knows very well, it was the seventeenth-century philosopher René Descartes who laid the philosophic foundations of modern physics. Descartes conceived of the external or objective world as made up of so-called *res extensae*, extended things bereft of sensible qualities, which can be fully described in purely quantitative or

---

3. *The Emperor's New Mind* (Oxford University Press, 1989), p. 256.
4. Ibid., p. 296.

mathematical terms. Besides *res extensae* he posits also *res cogitantes* or thinking entities, and it is to these that he consigned the sensible qualities, along with whatever else in the universe might be recalcitrant to mathematical definition. One generally regards this Cartesian partition of reality into *res extensae* and *res cogitantes* as simply an affirmation of the mind-body dichotomy, forgetting that it is much more than that; for not only has Descartes distinguished sharply between mind and body, but he has at the same time imposed an exceedingly strange and indeed problematic conception of corporeal nature, a conception, in fact, that renders the external world unperceived and unperceivable. According to René Descartes, the red apple we perceive exists—not in the external world, as mankind had believed all along—but in the mind, the *res cogitans*; in short, it is a mental phantasm which we have naïvely mistaken for an external entity. Descartes admits, of course, that in normal sense perception the phantasm is causally related to an external object, a *res extensa*; but the fact remains that it is not the *res extensa*, but the phantasm that is actually perceived. What was previously conceived as a single object—and what in daily life is invariably regarded as such—has now been split in two; as Whitehead has put it: "Thus there would be two natures, one is the conjecture and the other is the dream."[5] Now, this splitting of the object into a "conjecture" and a "dream" is what Whitehead terms "bifurcation"; and this, it turns out, constitutes the decisive philosophic postulate which underlies and determines our customary interpretation of physics. Beginning with his Tarner Lectures (delivered at Cambridge University in 1919), Whitehead has insistently pointed out and commented upon this fact. "The result," he declared, "is a complete muddle in scientific thought, in philosophic cosmology, and in epistemology. But any doctrine which does not implicitly presuppose this point of view is assailed as unintelligible."[6] I am here to tell you that today, after seventy years of quantum debate, the situation remains fundamentally unchanged. Just about every other article of philosophic belief, it would seem, has been put on the table and subjected to

5. *The Concept of Nature* (Cambridge University Press, 1964), p. 30.
6. *Nature and Life*, op. cit., p. 6.

ontology; but let us continue. Not only is God's love the unfathomable source of all causality, but all causation, as we know it, *imitates* that love. To quote Gilson once more:

> Beneath each natural form lies hidden a desire to imitate by means of action the creative fecundity and pure actuality of God. This desire is quite unconscious in the domain of bodies; but it is that same straining towards God which, with intelligence and will, will blossom forth into human morality. Thus, if a physics of bodies exists, it is because there exists first a mystical theology of the divine life. The natural laws of motion, and its communication from being to being, imitate the primitive creative effusion from God. The efficacy of second causes is but the counterpart of His fecundity.[10]

This same Thomistic vision of Nature has been expressed by Meister Eckhart in a passage of rare beauty which I would like also to share with you, in which he writes:

> You must understand that all creatures are by nature endeavoring to be like God. The heavens would not revolve unless they followed on the track of God or of his likeness. If God were not in all things, Nature would stop dead, not working and not wanting; for whether thou like it or no, whether thou know it or not, Nature fundamentally is seeking, though obscurely, and tending towards God. No man in his extremity of thirst but would refuse the proffered draught in which there was no God. Nature's quarry is not meat or drink . . . nor any things at all wherein is naught of God, but covertly she seeks and ever more hotly she pursues the trail of God therein.[11]

Here we have it: a vision of Nature which penetrates to the very heart of things, to that "most profound element" in fact, which St. Thomas has identified as its act-of-being. And to be sure, this is no longer an Aristotelian, but an authentically Christian *Weltanschauung*. I propose to show next how the findings of quantum theory fit into that Christian worldview.

---

10. Ibid., p.184.
11. *Meister Eckhart,* C. de B. Evans, trans. (London: Watkins, 1925), vol. 1, p.115.

It needs to be pointed out, first of all, that the Thomistic philosophy, no less than the Aristotelian, is unequivocally nonbifurcationist. There is not the slightest trace of "Cartesian doubt" to be found in either philosophy. What we know by way of sense perception are external objects, period; and these are the objects that enter into the Thomistic ontology. It follows that the findings of physics (our physics, that is) can be assimilated into the Thomistic worldview only on condition that they be first interpreted in nonbifurcationist terms.

The fundamental problem, clearly, is to situate the physical domain ontologically in relation to the corporeal. Now, we know that transitions from the physical to the corporeal are effected by acts of measurement in which a certain possibility inherent in a given physical system is actualized; and this constitutes, Thomistically speaking, a passage from potency to act. A physical system as such may consequently be conceived as a potency in relation to the corporeal domain. And I might add that this point has in fact been made forcefully by Heisenberg with reference to fundamental particles: "a strange kind of physical entity just in the middle between possibility and reality"[12] he calls these entities, and goes on to observe that in certain respects they are reminiscent of what he terms "Aristotelian *potentiae*." However, when it comes to the macroscopic domain, that is to say, to aggregates of fundamental particles that constitute the SX of a corporeal object X, Heisenberg identifies X and SX without the least scruple—as if the mere aggregation of atoms could effect a transition from potency to act! Nonbifurcation, on the other hand, implies, as we have seen, an ontological distinction between X and SX, which is to say that SX, no less than the quantum particles out of which it is composed, constitutes itself "a strange kind of physical entity just in the middle between possibility and reality." To be precise, fundamental particles and their aggregates—be they ever so macroscopic!—occupy a position, ontologically speaking, between primary matter and the

12. *Physics and Philosophy* (New York: Harper & Row, 1962), p. 41.

corporeal domain. It appears that contemporary physics has discovered an intermediary level of existence unknown and undreamt of in pre-modern times: and this is what I term the "physical universe."

What is it, then, that differentiates the two ontological planes? From an Aristotelian or Thomistic point of view the answer is clear: what distinguishes a corporeal object X from SX is precisely its substantial form. It is this form that bestows upon X its corporeal nature and specific essence, its "whatness" or *Sosein*, as we have said. And it is important to emphasize that this substantial form is perforce something other than a mathematical structure; for indeed, if it were, X and SX would in fact coincide. One might say that SX itself comprises everything in X that is "quantitative" or reducible to mathematical structure in the widest sense. Substantial forms, therefore, are not amenable to a quantitative or rigorously mathematical science. It is to be noted, moreover, that this fact was clearly recognized by Descartes himself: "We can easily conceive," he writes, "how the motion of one body can be caused by that of another, and diversified by the size, figure and situation of its parts, but we are wholly unable to conceive how these same things can produce something else of a nature entirely different from themselves, as for example, those substantial forms and real qualities which many philosophers suppose to be in bodies."[13] It needs however to be noted that this is precisely the reason why the protagonists of universal mechanism, headed by Galileo and Descartes, rejected substantial forms and banished sensible qualities from the external world: substantial forms as well as sensible qualities had to be excluded because neither could be reduced to mechanical terms. In so doing, however, Galileo and Descartes have cast out the very essence of corporeal being; one is left with a de-essentialized universe, a world emptied of reality.

We need today to free ourselves from the iron grip of this spurious and dehumanizing dogma. We need to rediscover the fullness of the corporeal world, replete with substantial forms and real qualities, which moreover enshrines at its very core the mystery of what

---

13. See E.A. Burtt, *The Metaphysical Foundations of Modern Physical Science* (New York: Humanities Press, 1951), p.112.

St. Thomas calls "the most profound element in all things." We have need of this discovery in every domain of life, including the scientific: even when it comes to the philosophic or "more-than-operational" understanding of quantum theory, as we have seen. But we have need of a sound ontology above all in the spiritual domain: authentic Christianity, in particular, demands a sacramental capacity on the part of matter which is categorically inconceivable in Cartesian terms. It hardly needs pointing out that in a universe comprised of quantum particles—in which not even a red apple can exist!—the Christic mysteries have absolutely no place. Now, I surmise that of all the true philosophies—and I believe there may be more than one—the Thomistic is for us the safest and most efficacious means by which to effect the liberating intellectual rectification. Whosoever has sensed that "love is the unfathomable source of all causation" has already broken the chains; and whoever has grasped, even dimly, what St. Thomas terms the "act-of-being" is well on his way.

2

Eddington and the
Primacy of the Corporeal

In his Tarner Lectures of 1938, published as *The Philosophy of Physical Science*, Sir Arthur Eddington has enunciated reflections on the nature of physics which to this day challenge our understanding of the physical universe. As is well known, Eddington championed a subjectivist interpretation of physical science: "selective subjectivism," he called it. Comparing the physicist to a fisherman catching fish with a net, he contends that the known laws of physics can be deduced simply from the structure of the "net" itself. Even the basic numerical "constants of nature," as they are called, can supposedly be elicited by epistemological considerations alone. In a technical treatise, entitled *Relativity Theory of Protons and Electrons* (published in 1936), he claims in fact to have accomplished these very feats of mathematical derivation. One is bound to wonder, of course, whether there might not be some gap or hidden flaw in these putative deductions. Suffice it to say that the significance and lasting relevance of Eddington's philosophy does not hinge on this question. It is no doubt true that his outlook, though widely respected, has rarely met with full belief: how could it be otherwise when someone, no matter how illustrious, purports to deduce—without recourse to a single experiment—that there are exactly 136×2^{256} electrons in the universe![1] I would however add that the more carefully one reads *The Philosophy of Physical Science*, the less fantastic such claims will seem.

---

1. Positrons are counted negatively, so that the total remains unchanged by electron-positron interactions.

Meanwhile it happens that the subjectivist interpretation of physics has, in any case, been gaining ground as a major trend, driven by the evolution of physics itself. In light of quantum theory one has come to understand that the act of measurement affects the system under observation in ways that can be neither predicted nor controlled. The presumed objectivity of physics has thus been compromised, and one hardly knows where to draw the line: how much, in other words, of what the physicist "discovers" may actually be the result of his own intervention. The very conception of physics has changed; as Heisenberg has put it: "Science [meaning physics] no longer stands before Nature as an onlooker, but recognizes itself as part of this interplay between Man and Nature."[2] This "interplay," moreover, is evidently effected by the act of measurement: it is here that the physicist acts upon Nature, and it is here too that Nature responds. Could it be, then, that Eddington was right after all: that the "net" of mensuration does determine the laws—and perhaps even the universal constants—of physics? It appears, on the strength of recent findings by an American physicist named Roy Frieden, that this is indeed the case. In a monograph entitled *Physics from Fisher Information*, published by Cambridge University Press in 1998, Frieden has essentially done what Eddington declared to be doable: from an analysis of the measuring process, Frieden has deduced the corresponding laws of physics. The analysis is information-theoretic, and Frieden employs a little-known information functional, discovered in 1925 by a statistician named Ronald Fisher. What takes place, according to Frieden, in the input space of a measuring instrument, is a transfer of Fisher information from an information content J, "bound" to the phenomenon, to the acquired information content I from which the data are sampled. Frieden's great discovery is that the information difference I minus J (the so-called *physical* information K) satisfies a variational principle: roughly speaking, Nature contrives to minimize K.[3] The decisive

---

2. *Das Naturbild der heutigen Physik* (Hamburg: Rowohlt, 1955), p. 21.

3. Frieden shows that this "interplay" between Nature and the instrument of mensuration can be conceived as a so-called "zero-sum" game between the physicist

fact is that the physical information K turns out to be a universal Lagrangian from which the laws of physics can be derived.

The resultant approach—dubbed "the method of extreme physical information" or EPI for short[4]—has been successfully applied to the major domains, and no end of its scope is yet in sight. As a matter of fact, physicists seem for the most part to be interested in EPI, not so much on account of its enormous philosophical implications, but mainly because it is proving to be a powerful research tool. For instance, EPI is presently being applied to problems in quantum gravity and turbulence which have proved recalcitrant to other means of approach.

What presently concerns us is the fact that Frieden's spectacular results go a long way in support of Eddington's philosophy. Yes, it does appear to be the interaction between the measuring instrument and the measured system—the "net" and the "fish"—that accounts for the laws of physics, as the British savant had long ago foretold.[5]

I propose now to present a summary of Eddington's philosophy, followed by critical reflections consonant with the Aristotelian and Thomistic traditions. I shall argue, in particular, that the very logic of Eddington's approach *demands* the distinction between what I have termed the *physical* and the *corporeal* domains, and moreover entails the ontological primacy of the corporeal.

seeking information and an opponent reluctant to part with it. It happens that the opponent (representing Nature) generally wins.

4. I should note that EPI imposes one further condition: in addition to the stated variational principle it imposes an algebraic condition on I and J. Thus amended, EPI constitutes a "top law" from which most if not all the laws of physics can be derived. The EPI principle itself, however, though arrived at by way of heuristic considerations, is not deduced *à la* Eddington through epistemological analysis. Frieden's claims are thus a bit more modest.

5. This is not to say that there must be some mathematical kinship between Frieden's EPI approach and Eddington's group-theoretic derivations. The only thing these exceedingly different derivations have in common—on the surface, at least—is the amazing claim to deduce the fundamental laws of physics from an analysis of the measuring process.

Eddington commences his inquiry into the nature of physics with an appropriate definition of the physical universe: "Physical knowledge (as accepted and formulated today)," he writes, "has the form of a description of a world. We *define* the physical universe to be the world so described." (3)[6] The physical universe, according to this definition, is not simply the universe as such, but the universe "so described": it is the world as seen through the lenses of the physicist. And these "lenses," Eddington contends, introduce subjective elements into the physical universe, i.e., the universe "so described." Certainly the descriptions at which the physicist arrives are not wholly subjective; yet Eddington insists that the *laws* of physics, as distinguished from what he terms special facts, *are* wholly subjective. And that is precisely the reason why these laws can indeed be known with mathematical precision: the vaunted precision of physics, Eddington declares, derives in fact from its subjectivity; as Whitehead once put it: "Exactness is a fake." The so-called special facts, on the other hand, are objective to some extent, "depending partly on our procedure in obtaining observational knowledge, and partly on what there is to observe." (66) Take for instance the fact that the Moon is so and so many miles distant from the Earth: although this finding presupposes evidently an observational procedure for measuring distance, it is not determined by that procedure alone: not even Eddington would go that far! The aforesaid assertion concerning the Moon has thus an objective content; but for that very reason the distance in question turns out not to be knowable with absolute precision. We will return to that side of the story later, in connection with quantum theory and the role of probabilities.

Having distinguished between laws and special facts, one finds the latter to be the more problematic, on account of their "partly objective, partly subjective" character. What about the structure of

6. The page numbers in parentheses refer to *The Philosophy of Physical Science* (Cambridge University Press, 1939).

covered by the physicist are simply there, even as Mt. Everest, say, is simply there; but Eddington disagrees:

> Theoretical physics is highly mathematical. Where does the mathematics come from? I cannot accept Jeans' view that mathematical conceptions appear in physics because it deals with a universe created by a Pure Mathematician: my opinion of pure mathematicians, though respectful, is not so exalted as that. (137)

Eddington's point is that "The mathematics is not there till we put it there." Moreover, we begin to "put it there" the moment we define the most basic of all measures: the measures of spatial distance, namely, and of temporal duration. According to Eddington's analysis, this step entails the mathematical theory of groups, which consequently stands at the base of our physics. It is by means of mathematical groups, constituted by what Eddington calls "terminable sets of operations" (140), that mathematics gains a foothold, so to speak, on the ground of external reality.[9] Or better said: on the ground of human knowledge; for as we have seen, it is Eddington's contention that the mathematics pertains, not to the world as such, but to our *knowledge* of the world. It is in a sense subjective: we ourselves have put it there. And this means that the physical universe—the universe "so described"—is likewise subjective in precisely the same sense: i.e., because it is constituted by mathematical structures which the physicist himself has imposed. This, then, is Eddington's central point, and ultimately his *only* point: the conclusion to which all his arguments lead.

But let us continue. Starting with terminable sets of operations, one arrives at the abstract mathematical group, in which the original operations are represented by symbols, such as the x, y, and z of algebra, which now function as undefined elements. What counts—and this is indeed the crux of the matter!—are not of course these symbols, but the mathematical structure represented by their means: it is this structure, and this alone, that has physical significance. As Eddington clearly explains:

---

9. What thus enters into the foundations of physics are certain special groups, such as the group of rotations in six-dimensional space.

Physical science consists of purely structural knowledge, so that we know only the structure of the universe which it describes. This is not a conjecture as to the nature of physical knowledge; it is precisely what physical knowledge as formulated in present-day theory states itself to be. In fundamental investigations the conception of group-structure appears quite explicitly as the starting point; and nowhere in the subsequent development do we admit material not derived from group-structure. The fact that structural knowledge can be detached from knowledge of entities forming the structure gets over the difficulty of understanding how it is possible to conceive a knowledge of anything which is not part of our own minds. So long as the knowledge is confined to assertions of structure, it is not tied down to any particular realm of content. (142)

It might seem, at first glance. that Eddington has simply enunciated the Aristotelian and Thomistic doctrine of knowledge: it is by way of a form that we know an external object. But one must remember that the knowledge of which Aristotle and Aquinas speak is *objective*: we know the very form that resides in the object (be it a substantial or an accidental form). Why, then, is the knowledge of mathematical forms or structures said to be subjective? And does this not contradict Eddington's claim that it is precisely by way of mathematical forms that the physicist *is* able "to get out of his mind" and make contact with the external world? There is however no contradiction here. Yes, it is indeed by way of the mathematical forms that the physicist gains knowledge of the external world; Eddington's point, however, is that the forms in question have been artificially imposed: "The mathematics is not there until we put it there." And it is for this reason, and in this sense, that our knowledge of mathematical structures—our knowledge of the physical world!—is said to be subjective.

The so-called physical universe—"the world so described"—turns out to be constituted by mathematical structures which we ourselves have imposed; in a word, it proves to be "man-made." Yet this fore be indeterminate, yet capable of receiving determination. Not "fully indeterminate," to be sure; the latter notion would be as chimerical as that of a classical particle, which is its logical opposite. The measurable, thus, must be situated between these two extremes. Though reminiscent in certain respects of Aristotle's *potentiae*, it is more than that; for the measurable in the sense of physics carries a certain tendency, which in fact can be conceived in mathematical terms: as a *probability*, to be exact. The measurable, one finds, is thus in effect a probability. It is by way of probabilities, it turns out, that Nature outwits our two-valued logic of "being" and "nothingness": for indeed, a probability is neither the one nor the other, but does evidently occupy a middle ground.

The notion is philosophically difficult, and even Eddington seems to vacillate in regard to what these probabilities are. At one point, for example, he tells us that "The introduction of probability into physical theories emphasizes the fact that it is knowledge that is being treated. For probability is an attribute of our knowledge of an event; it does not belong to the event itself, which must certainly occur or not occur." (50) Yet, if probability is indeed an attribute of our knowledge, it should tell us something about the object or system that is known. According to quantum theory, in fact, probabilities are all that we do know in the physical world.[16] As Eddington points out: "Wave mechanics investigates the way in which probability redistributes itself as time elapses; it analyses it into waves and determines the laws of propagation of these waves." (51) One is beginning to think that these waves are real! But almost immediately he adds that "a sudden accession of knowledge—our becoming aware of a new observation—is a discontinuity in the 'world' of probability waves," suggesting once again that these probability waves pertain to our knowledge alone—as if they did not, by that very fact, belong also to the physical world. It needs to be understood, moreover, that the discontinuity in question arises not simply from an "accession of knowledge—our becoming aware of a

---

16. Strictly speaking, "all that we do know" are *data values*, as Professor Frieden has kindly pointed out. Probabilities, thus, are not "known," but inferred. They are "all that we *can* infer."

new observation," but from that new observation itself, whether we become aware of it or not. What counts is the interaction of the measuring instrument with the physical system that is being measured: it is this that causes the so-called collapse of the wave function. Some forty pages later, however, it appears that Eddington adopts at last an unequivocally realist interpretation of probabilities: "I must still keep hammering at the question, What do we really observe?" he writes. "Relativity theory has returned one answer—we only observe *relations*. Quantum theory returns another answer—we only observe *probabilities*." Here we have it: what we "really observe" are probabilities. The circumstance, moreover, that relativity theory gives a different answer does not alter this fact: Eddington clearly accepts quantum theory as the more accurate and indeed the more fundamental of the two theories. The conclusion stands: *probabilities are "what we really observe."*

But if there are indeed probabilities, there must also be things which are *not* probabilities; a probability is after all the probability of something which itself is not a probability. In addition to probabilities, thus, there are also events "which must certainly occur or not occur," to put it in Eddington's words. A distinction needs therefore to be drawn between two kinds of facts; as Eddington explains:

> Probability is commonly regarded as the antithesis of fact; we say 'This is only a probability, and must not be taken as a fact.' What we mean is that the result of an observation, though undoubtedly a fact in itself, is only valuable scientifically because it informs us of the probability of some other fact. These secondary facts, known to us only through probabilities, form the material to which the generalizations of physics refer. (89)

We must not fail to observe that in this somewhat obscure passage the distinction between the corporeal and the physical domains lies concealed. Eddington alludes to two kinds of facts: the result of an observation, namely, and the so-called "secondary facts, known to us only through probabilities," which are said to "form the material to which the generalizations of physics refer." Now the result of a measurement, as we have seen, is realized in the perceptible state of a corporeal instrument: it is thus a "corporeal" fact, the kind which are not probabilities. The "secondary facts," on the other hand, refer evidently to the physical domain, the domain in which facts *are* probabilities, or are "known to us only through probabilities," which amounts to the same.

The realization that "we only observe probabilities" proves to be basic; and I would point out that Frieden's EPI approach substantiates this most forcefully. The carrier of Fisher information, after all, is none other than a probability distribution. There is an input probability distribution, corresponding to the "bound" information J, and an acquired probability distribution, corresponding to I, from which the measured value is sampled; it is here that probability collapses, so to speak, into fact (fact of the first kind, in Eddington's enumeration). Up to that final point—which is moreover impenetrable to mathematical analysis—the measuring process has to do with probabilities: "The basic elements of EPI theory," writes Frieden, "are probability amplitudes."[17] What takes place in measurement is a transmission of probabilities, which gives rise to a corresponding transmission of Fisher information from J to I; and it is Frieden's monumental discovery that this very process determines the physics of the measured phenomenon in the form of an output law. As Frieden explains: "EPI regards each such equation as describing a kind of 'quantum mechanics' for the particular phenomenon."[18] This holds true, moreover, even when one derives seemingly "non-quantum" results via EPI: the classical electromagnetic four-potential, for instance, can now be regarded as a probability amplitude for photons; or again, the gravitational metric tensor in Einstein's general theory becomes a probability amplitude for so-called gravitons. According to EPI, *all of physics* is a quantum theory; how could it be otherwise if "we only observe probabilities"?

What then, let us ask, *is* a probability? Now, the defining characteristic of a probability, clearly, is that it is the probability of something which itself is *not* a probability. And when it comes to the probabilities of quantum theory, that "something" is of course none other than the result of a measurement, as given, say, by the posi-

17. Op. cit., p.83.
18. Ibid., p.84.

tion of a pointer on a scale. It is here, in this transition, that a probability reveals its nature; and let us not fail to observe that this very transition—which in the standard formulation of quantum theory reveals itself in the phenomenon of wave function collapse—presents itself as an indeterminacy.

Let us reflect upon this fact, which proves to be of the utmost importance. If probabilities are what we really observe, they must be at least "partly objective"; but then, if it be true that transcendence is indeed "the hallmark of objectivity" (as we have said), must they not reveal that partial objectivity in some mark of transcendence? Now, that mark of transcendence, I say, stares us in the face: it is obviously none other than the so-called quantum indeterminacy of wave function collapse. By way of this crucial enigma Nature contrives to outwit our simple logic: by this physically inexplicable "collapse" into objective fact the probabilities of physics manifest their transcendence: at this very moment they reveal their objective side by violating the Schrödinger wave equation, which up to that point they had strictly obeyed.[19] In a "flicker of transcendence" as it were, they announce their partial objectivity, and thus their partial subjectivity as well: for the statement itself is made, not on the physical plane of probabilities, but on the plane of *corporeal* fact, that is to say, in a perceptible state of a corporeal instrument. The entire mystery of the physical is in a way revealed in the spectacle of wave function collapse: no wonder the phenomenon has been pondered and debated since the advent of quantum theory! And no wonder that it continues to baffle the serious-minded physicist: for it happens that the probabilities of quantum physics announce their objectivity precisely by *violating* the quantum-mechanical equations. In a momentary burst of non-compliance, one could say, they prove to be *more* than a mathematical abstraction, more than a mere *ens rationis*, more in fact than just an imprint of Eddington's "net." By their momentary breach of "exactness" the probabilities of physics attest that they are *not* "a fake."

---

19. According to the EPI approach, the Schrödinger wave equation is not violated but "re-initialized" as Professor Frieden points out, which however amounts to the same.

It behooves us now to return to Eddington's "two tables" to see where the matter finally stands. We have taken the pre-Cartesian and indeed "normal" position, which affirms that the perceived table is situated, not "in consciousness," but in the external world: it is the real table, we contend, the one and only table that objectively exists. But where exactly does this leave the physical table: the one composed of atoms or particles? We have rejected the idea that Eddington's two tables have simply changed places, so that it is now the molecular table instead of the perceived that is relegated to the realm of consciousness, the Cartesian *res cogitans*. But this means that *both* tables are now situated in the external world: how can that be? Could the second be conceived perhaps as a kind of ghost-like double pervading the corporeal imperceptibly? That would be an "embellishment" tantamount to corporealizing the physical. No matter how thin or ethereal we picture that "ghost" to be, it remains perforce a corporeal entity; but "things physical," it turns out, are *not* corporeal, nor are they, strictly speaking, "things": the fact that they are characterized in terms of *probabilities* should suffice to make this clear. The idea of the "ghost-like double" needs therefore to be rejected.

What is actually observed, we have said, are the probabilities of physics. They constitute the objective reality which underlies modern physics, the bedrock upon which the edifice of modern physics rests. As that which is actually observed, they—and they alone!— cannot be assigned to a model or representation. But neither, as we have seen, can they be assigned to the objective universe, which proves to be none other than the corporeal: as Eddington points out, the probabilities of physics are indeed facts, but facts "of the second kind" inasmuch as they cannot be separated from the economy of measurement. They are thus "partly subjective" as we have said, or facts of the "participatory universe" to put it in John Wheeler's felicitous phrase.[20]

---

20. "All things physical," writes the Princeton physicist John Wheeler, "are information-theoretic in origin, and this is a participatory universe." This remarkable statement is (understandably!) quoted on page 1 of Frieden's book.

It appears that the enterprise of physics has given rise to a new ontological domain, a new stratum or substratum of reality. By virtue of its "participatory" nature it turns out however that the physicist himself determines the contours of that so-called universe. What then, let us ask, is its ontological status? The answer has by now become clear: *the physical stands to the corporeal as potency to act*, to put it in Scholastic terms. This is what the aforesaid Eddingtonian considerations force us to conclude: the recognition that "what is actually observed" are finally *probabilities* implies as much. What Heisenberg affirmed with reference to fundamental particles applies consequently to the physical universe at large: stripped of its imaginary "embellishments," it too is "a strange new entity midway between possibility and reality." Metaphysically speaking, that so-called universe constitutes a *materia secunda* situated between *materia prima* or pure potency and the corporeal domain.[21]

21. For a fuller discussion of these issues I refer, in the first place, to my monograph, *The Quantum Enigma,* op. cit. See also *Science and Myth,* op. cit., Chapters 2 & 3.

3

The Ontology of Bell's Theorem

In the present chapter I propose to examine the quantum-mechanical notion of "non-locality" in light of traditional ontology. The text is based upon a lecture delivered in 1988, in which, perhaps for the first time, an aspect of the quantum-reality problem was dealt with from a metaphysical point of view.

I will begin by recalling certain well-known quantum facts which came into scientific view near the turn of the century, precisely at the moment when a victorious physics seemed to be closing in upon the long-conjectured atoms and the fundamental particles of which they now proved to be made. What happened in that final moment, so to speak, is that the quarry has mysteriously eluded our grasp: for in light of experimental scrutiny these so-called fundamental particles exhibited a hitherto unsurmised wave aspect, implying that they were not, strictly speaking, particles at all. The underlying reality, therefore, proved to be an unknown, an X, which under appropriate conditions can exhibit both particle and wave characteristics, and hence cannot be conceived in classical terms. What this X is in its own right, no one knows; the quarry, as I have said, has quite eluded our grasp.

To be sure, what is actually observed—what registers on our instruments of detection as a condensation track, or a flash of light on a phosphor screen, or as a spot on a photographic plate—is indeed the particle aspect, or the particle as we continue to say. The wave aspect, on the other hand, manifests itself indirectly through statistical laws governing the observed behavior of the so-called particles. For instance, if a beam of electrons is passed through a small round aperture, what appears on the phosphor screen, or on the photographic plate, is a pattern of concentric bands (resembling

55

an archery target). Now, this is none other than the so-called Airy pattern, named after the British astronomer George Biddell Airy, who demonstrated, back in 1835, that such concentric bands can be explained mathematically in terms of the diffraction and interference of an impinging wave. Thus, in the statistical distribution of the individual electron encounters with the screen or plate, the electron beam acts as if it were a wave. Oddly enough, moreover, one gets exactly the same Airy distribution whether the electrons are passed through the aperture in vast numbers or sparsely, even one electron at a time. A single electron, therefore, has both a wave and a particle aspect. And whereas the particle aspect suggests that the X in question is discrete and localized, the wave aspect entails that it must be continuous and somehow spread out over the entire space of the experiment.

Such, in brief, are the quantum facts which brought the Newtonian age to an abrupt end. It quickly became apparent that the classical formalism of mathematical physics was inherently incapable of coping with the phenomenology of the newly-revealed quantum world; and by 1925, following a quarter century of theoretical chaos, a new and radically different fundamental theory came to birth. And this "quantum theory," as it came to be called, has proved to be a brilliant success. In its application to countless problems it has never yet failed, never yet yielded an incorrect response. It is truly a marvel, a scientific triumph.

However, there is one great problem with this marvelous theory, which is that by its very form it gives us no information—no inkling, even—concerning the nature of physical reality as such. The theory tells us, with uncanny precision, what our instruments of detection will register when a beam of electrons, for example, is subjected to certain conditions; but it tells us nothing whatsoever concerning the electron as such. Speculations aside, we do not actually know, for instance, whether the electron owns its dynamic attributes (such as position or momentum) *before* it is observed— whether it is thus what physicist term an "ordinary object," or whether these dynamic attributes are in fact "contextual" (which would mean that they exist *only* in the context of an actual measurement). It is basically the old conundrum whether external objects exist when there is no one present to perceive them—but with a very modern twist: for indeed, the problem today is whether it is actually possible to conceive of a noumenal quantum reality in such a way as not to contradict what we know about phenomena.

This is the issue which stands at the heart of the so-called quantum-reality problem, on which physicists have been sharply divided. As far back as 1927, when the heavyweights of physics gathered in Brussels to assess what, if anything, was still left of the Newtonian *Weltanschauung*, Niels Bohr and Albert Einstein took up diametrically opposing positions on this issue, which have polarized the debate ever since. What Bohr maintained, basically, is that in regard to quantum phenomena we must be satisfied with the pragmatic kind of information quantum theory itself delivers: it is the best one can do, not because the theory as such is imperfect or incomplete (as Einstein charged), but because there is in fact no quantum reality *behind* the phenomena which could explain what is going on. "There is no quantum world," said Bohr. "There is only an abstract quantum description."

Einstein, on the other hand, argued with all his considerable might in favor of a realist interpretation. There *is* a "quantum world," he said in effect, a deep reality which underlies the phenomena and ultimately explains them: there must be! It was unthinkable to him that an "abstract quantum description," unsupported by any deep reality, could lead to correct predictions—as if by magic.

As it turned out, most physicists eventually joined the Bohr school of thought, which came to be known as the Copenhagen interpretation of quantum theory. In 1932, moreover, just five years after the Brussels conference, it seemed very much as though the Copenhagenist claims had been vindicated, once and for all, through the labors of a Hungarian mathematician by the name of John von Neumann. Having put the new quantum theory on a firm mathematical foundation, von Neumann went on to prove—with exemplary rigor!—that the idea of a physical reality made up of "ordinary objects" is in fact incompatible with the predictions of quantum theory. At this point it appeared to all but a few realist diehards—led by the unconquerable Albert Einstein—that the issue was henceforth closed.

But in fact it was not. The realist position revived, phoenix-like, in 1952, when David Bohm, after extensive conversations with Einstein, succeeded in constructing an objective model of the electron which squared with the exacting demands of quantum theory. In the wake of the celebrated "von Neumann proof," Bohm had apparently accomplished the impossible. It thus became apparent that there must be an error, a loophole of some kind, in von Neumann's argument; and yet the great mathematician had done his work so well that it took another twelve years to find it. The riddle was finally solved in 1964 when a hitherto unknown physicist by the name of John Stewart Bell discovered that von Neumann had unwittingly made a hidden and indeed unwarranted assumption concerning the "ordinary objects" which quantum theory supposedly had ruled out. What von Neumann had presupposed—and what until then everyone had apparently assumed to be self-evident—was that the objects in question were "local" entities. Basically this means that they can only communicate with each other via known physical forces: via signals, therefore, which cannot propagate faster than the speed of light. It turns out that Bohm was able to obviate the von Neumann interdict precisely because his model of the electron did not in fact obey that stipulated condition: Bohm's electron happens not to be a *local* entity.

But this result—this decisive breakthrough!—proved to be only the first step in the unfolding of Bell's inquiry. Basing himself on the fundamental principles of quantum theory, he succeeded in proving that reality, be it ordinary or contextual, *must* in fact violate von Neumann's locality condition. It thus turns out that the locality postulate is not only unwarranted, but proves actually to be untenable: in a word, *reality is nonlocal.* This is the epochal discovery which has come to be known as Bell's theorem. It has since been verified experimentally on the basis of an inequality likewise discovered by Bell, which moreover does not hinge upon quantum theory; the result, therefore, appears to stand solid as a rock. But whereas most physicists today accept Bell's theorem as well-founded, a few continue to be skeptical, and have left no stone unturned in an effort to thwart Bell's conclusion. And perhaps rightly so, considering that *de jure* the affirmation of nonlocality revolutionizes our worldview: it is not without reason that Berkeley physicist Henry Stapp refers to Bell's theorem as "the most profound discovery of science."[1]

What makes it so is the fact that its conclusion points beyond the spatio-temporal universe, the domain to which, strictly speaking, physics is confined by the very nature of its *modus operandi*. The astounding fact is that, in the form of Bell's theorem, physics has been forced to admit its own boundedness, its incapacity to deal with the deeper strata of cosmic reality. Having, for centuries, claimed title to the objective universe in its entirety, physics has now been forced, on its own ground, to relinquish that cherished pretension.

Actually, the Copenhagenist disclaimer amounts to the same; the die, one can see in retrospect, had been cast with the discovery of quantum mechanics. What Bell did was to isolate and bring into the open a fundamental feature of quantum theory which had engaged physicists from the start: the existence, namely, and indeed the ubiquity, of what are nowadays termed "nonlocal" connections. And that is a trait Einstein had in fact singled out as a fatal flaw of the new physics. Meanwhile however that seeming flaw has proved to be a fundamental truth indigenous to the quantum realm, an epochal discovery, in fact, which has revolutionized our very conception of physics. Yet, remarkably, it was by way of Einstein's "thought experiment" designed to discredit quantum theory that the aforesaid breakthrough was finally achieved. What ultimately stands at issue, one can now see, is whether or not the Einsteinian space-time constitutes the full locus of cosmic reality, as Einstein himself supposed. It happens that quantum theory gives answer to this question, and that answer proves to be negative. What clinches the matter, finally, is that the issue can be tested experimentally thanks to Bell's inequality, and the verdict is now in: on this point it is Einstein who has missed the mark. Contrary to what the great physicist believed, it happens that space-time itself is bounded; and this we now know on the strength of physics itself. As we have said, physics has at last declared its own incapacity to deal with the deeper strata of cosmic reality. Its domain has thus shrunk, as it

1. "Bell's Theorem and World Process," *Il Nuovo Cimento*, 40B (1977), p.271.

were, from the universe at large to a restricted class of phenomena; as one unusually astute and forthright quantum theorist has put it: "One of the best-kept secrets of science is that physicists have lost their grip on reality."[2]

What is in fact required if one would "grasp reality" are certain fundamental conceptions pertaining to the traditional metaphysical wisdom of mankind, which I would like now to place before you. We will take as our starting point the opening verse of Genesis: *In principio creavit Deus caelum et terram.* First of all, it needs to be clearly understood that the "beginning" in which God is said to have created the world is *not* to be conceived in temporal terms. It is definitely not "a moment of time," be it six thousand or fifteen billion years ago; for as St. Augustine observes: "Beyond all doubt the world was not made *in* time, but *with* time."

Regarding the second half of the biblical verse, we need to observe that what God creates is "*heaven and earth*": not one thing, thus, but apparently *two*. Now that "heaven" and that "earth" may be taken to refer respectively to the spiritual and the material poles of creation, or more concretely, to a spiritual and a corporeal domain. The point, in any case, is not that God created two separate worlds—a heavenly and an earthly—but that the cosmos in its integrality comprises two complementary principles or components. After all, if "heaven" and "earth" are brought into existence in a single creative Act (as the biblical verse affirms), it stands to reason that the two must constitute a single creation, that in conjunction they are one thing.

At this point, however, we need to avail ourselves of another fundamental insight: there must also be an intermediary principle or realm—a *metaxy* in the Platonist sense—for as Plato was perhaps the first to observe: "That two things of themselves form a good union is impossible." So too St. Thomas Aquinas declares: "The order of reality is found to be such that it is impossible to reach one

2. Nick Herbert, *Quantum Reality* (Garden City: Doubleday, 1985), p. 15.

end from the other without passing through the middle." One finds, therefore, that the cosmos in its integrality must comprise *three* ontological levels or degrees: in ascending order these are the corporeal, the intermediary, and the spiritual. Two crucial observations are called for at this point. It needs first of all to be noted that this ternary defines a dimension—a fourth dimension if you will— which in metaphysical parlance is termed *vertical*. In a word, the creation, metaphysically conceived, comprises an "above," which moreover has obviously been lost in the contemporary *Weltanschauung*. This is the first point; and the second is this: what holds for the creation at large holds true likewise for man: he too proves to be tripartite, composed as he is of *corpus, anima,* and *spiritus*.[3] Man proves thus to be more than a mere part of the universe—incomparably more than an infinitesimal speck in the vastness of space-time!—but constitutes in truth a microcosm, a veritable universe in miniature. He too, thus, embodies a vertical dimension; whether he realizes it or not, he is half angelic and half earthly in his composition.

The creation is thus to be conceived as a hierarchy made up of three principal levels or tiers. The entire structure, moreover, can be most fittingly represented in the form of a symbolic circle, in which the center (or if you will, a central disc) represents the spiritual world, the circumference stands for the corporeal or "visible" universe, and the annular region in between represents the *metaxy* or intermediary domain. I would like to add that this symbolic circle—this veritable icon of the *cosmologia perennis*— was known to every major civilization. The great exception, of course, is our own: this profane post-medieval civilization, intellectually dominated by science, which has in effect reduced the cosmos to its lowest tier.

Having recognized the tripartite division of the integral cosmos, we need now to ask ourselves what it is that actually differentiates the intermediary from the corporeal plane. The answer to this question turns out to be simple, at least in principle: what distinguishes the corporeal domain from the intermediary is the fact that the former is subject to certain *quantitative* conditions or bounds from

---

3. This anthropology, which traces back to St. Paul, does not conflict with the more familiar *corpus/anima* anthropology rightly understood.

which the latter is exempt. These bounds, moreover, constitute the basic determinations which permit us to speak of space, time, and matter (or energy) in a precise quantitative sense. Think of a boundless expanse of water, let us say; and then suppose that a pot, having a certain size and shape, is immersed therein. Obviously the vessel divides what before was undivided, and introduces quantitative determinations which did not previously exist. We have here a simple model, if you will, a paradigm which enables us to understand, in the first place, that the substantial reality is one and the same on both ontological planes: water, after all, does not cease to be water when it is poured into a pot. Clearly, the "vessel" does not affect the substance, but only gives rise to certain quantitative attributes, "accidents" in the Aristotelian sense.

Now, it needs to be clearly understood that the primary determinations which characterize the corporeal plane of existence are by no means subjective, or man-made, but derive precisely from the Fiat of God. These primary bounds have been imposed, figuratively speaking, by the Creator and Architect of the universe when He "*set His compass upon the face of the deep*" as the Book of Proverbs has so beautifully expressed it. We need not concern ourselves with the question how, based upon the primary or God-given determinations, the *modus operandi* of physics gives rise to secondary quantitative determinations, which can then be described in mathematical terms. Suffice it to say that the quantitative determinations which constitute the immediate objects of modern physics presuppose the primary cosmic bounds which determine or define the corporeal plane of existence. What then are these primary bounds? It will suffice to say that they are none other than the determinations which underlie our pre-scientific notions of *space*, *time*, and *matter*.

When it comes to the intermediary domain, on the other hand, these customary notions do not apply, or at least do not apply in the accustomed sense. And this entails that our scientific measures of space, time, and mass or energy—which, as we have said, presuppose the former—become inapplicable as well. By comparison to the corporeal domain, the intermediary world is thus "unobstructed" in a manner which does indeed defy our ordinary human means of comprehension, be they sensory or scientific.

62

We are ready now to return to the problematic of quantum theory. "Everything we know about Nature," writes Henry Stapp, "is in accord with the idea that the fundamental process in Nature lies outside space-time but generates events that can be located in space-time."[4] Can it be said, one is now constrained to ask, that "the fundamental process in Nature," which proves not to be subject to the ordinary bounds of space and time, is in fact situated on the intermediary plane? Can it be said, thus, that physics, on its most fundamental level, has in a way *rediscovered* the intermediary world? One sees that such is "in a way" the case. Not, to be sure, in the sense of a discovery or theorem that can be formulated on the level of physics itself. What physics *can* prove, and what it has indeed established beyond reasonable doubt, is that external reality—and thus the cosmos as such—cannot be confined within the bounds of Einsteinian space-time: for if it could be thus confined, it would perforce satisfy Einstein's condition of locality, which in fact it does not obey. Inasmuch, however, as the external reality transcends the bounds of Einsteinian space-time, it escapes the grasp of contemporary physics; Nick Herbert, it turns out, has it exactly right: physicists *have* "lost their grip on reality." It needs however to be added that they never had such a "grip" in the first place. What physics deals with in the final count, to put it in terms of our erstwhile metaphor, is not the "water," but the "pot."

But why should that recognition amount to a "rediscovery of the intermediary world," as we have suggested? This is not of course a scientific question, but one that belongs incurably to the domain of traditional ontology. And the fact is that it does: to speak of "a process in Nature" outside of space-time as does the physicist Henry Stapp is most assuredly to situate that "process" on the intermediary plane.[5] What validates the inference is simply the fact that the things of the spatio-temporal world are none other than the things

---

4. "Are Superluminal Connections Necessary?" *Il Nuovo Cimento.* 40B (1977), p. 191.

5. One must remember that "above" the intermediary domain, the notion of "process" loses all sense.

of the intermediary realm *subjected to spatio-temporal bounds*: remove these bounds, and what remains pertains indeed to the intermediary plane. And I might mention in passing that the transition from the former to the latter is precisely the alchemical *solve*, even as the reverse transition constitutes the complementary *coagula*: like the reality itself, the operations of alchemy do not fit into the spatio-temporal stratum of the universe. Getting back to nonlocality: whatever the quantum physicist himself may think, he *has* in effect rediscovered the intermediary world.

The question remains of course what the physicist does think in that regard. Now to be sure, most physicists do not "think" at all: they simply accept the quantum facts and go about their business. One knows very well, moreover, what that business is: its object is to discover new facts of the same generic kind, and comprehend all these findings, so far as one can, within a single mathematical structure. The question of "being," of ontology—of *Weltanschauung* properly so called—does not arise at all; only a small minority of scientists, it appears, have seriously pursued these deeper issues. They are the quantum-reality theoreticians; one should however add that though their number be small, their ranks include most of the great founders, men of the stature of Bohr and Einstein, Heisenberg, Schrödinger and Planck. I have already referred to the fact that Einstein and Bohr assumed opposing positions, and that most quantum reality theorists eventually sided with Bohr. The majority came thus to believe that the pre-quantum worldview, which most physicists had tacitly assumed, needs now to be abandoned; and so began the search for a new philosophical *Ansatz* that could account for the paradoxical findings of quantum theory. A wide variety of philosophical conceptions have since been pressed into service, including notions that are truly bizarre, such as the various kinds of "many-worlds" theories presently in vogue; yet it appears that these labors have so far succeeded only in exacerbating the philosophical confusion. It would seem that just about every conceivable remedy has been tried, with a single exception: not one of these philosopher-scientists has placed himself on traditional metaphysical ground.

What is called for, as I have argued repeatedly, are certain speculative keys which the perennial wisdom alone can supply. And these keys are in fact none other than the principles of traditional cosmology which I have enumerated summarily in the Introduction to this book. In the preceding two chapters we have been primarily concerned with the first of these principles, which is tantamount to the tenet of non-bifurcation: it was the systematic application of this principle that led us to the ontological distinction between the physical and the corporeal domains, and has made possible the subsequent reinterpretation of physics as such. In the present chapter, which has to do with the quantum-mechanical finding of nonlocality, we have applied the second cosmological principle, which refers to the hierarchic structure of the integral cosmos. The key to the understanding of nonlocality, it turns out, is to be found in the traditional distinction between the corporeal and the intermediary levels of cosmic manifestation, and thus between the *sthūla* or "gross" and the *sūkshma* or "subtle" modes of being, to put it in Vedantic terms. As we have come to see, the "process in Nature" that gives rise to the observable quantum phenomena is in fact "subtle" in precisely the Vedantic sense, and is consequently situated, ontologically speaking, "above" the corporeal level. Strangely enough, quantum physics itself, when interpreted from a metaphysical point of view, distinguishes thus between *three* ontological planes: in ascending order, these are the physical, the corporeal, and the intermediary. To be precise, quantum mechanics affirms the first ontological discontinuity or hiatus through the phenomenon of state vector collapse, and the second through its recognition of nonlocality.

No wonder these two phenomena constitute the great enigmas of quantum theory! No wonder both should have mystified the physics community; for it happens that neither can be understood from the perspective of physics itself. Positioned, as he is, on the corporeal plane, the physicist is looking "downwards" at the physical. Ontologically speaking, his perspective is thus centrifugal, which is to say that his intellectual gaze is directed *away* from the ontological center, the pole of "being" and "essence." What ultimately confronts him, thus, is the outermost periphery of the integral cosmos, what the Scholastics termed the *materia secunda* or "container" of the universe, which has no being, no essence, no existence of its own. In this perspective he cannot see the corporeal, much less can he conceive of the intermediary realm, which is situated "above" the corporeal. To arrive at an understanding of what quantum theory itself has brought to light, he needs thus to undergo an intellectual conversion or *metanoia*: a revolution of 180° one can say.

Meanwhile the fact remains that quantum mechanics has "in its own way" rediscovered the intermediary domain. And this event is epochal. If the modern West is indeed the first society to view the corporeal world as a closed system, as Huston Smith observed, that error has now been detected on the basis of physics itself.

Needless to say, this recognition has decisive implications not only for philosophy, but for every fundamental domain of science as well. This holds true above all in the case of evolutionary biology, a discipline which in fact it disqualifies at one stroke. In view of the enormous importance of this question I will close with a brief consideration of that issue.

What Darwin failed to grasp, and what even most of his opponents have failed to realize, is that the corporeal universe in its entirety constitutes no more than the outer shell of the integral cosmos, and that the mystery of origins needs in fact to be resolved, not at the periphery, but precisely at the very center of that integral universe. So long as one imagines that the origin of a plant or animal may be conceived as a spatio-temporal event or process of some kind, one has missed the point. Briefly stated, corporeal beings have a double birth: a pre-temporal birth, first of all, in the divine creative Act, and a temporal birth which marks their entry into the corporeal domain. Now, what the first birth brings into being is what the Latin Fathers termed the *ratio seminale* and the Greeks the *logos spermatikos*, which is not however a corporeal entity—not actually a seed in the sense of biology—but a spiritual seed, one could say. That seed, metaphorically speaking, is sown at the Center, incubates in the intermediary domain, and is brought into manifestation on the corporeal at a moment of time "when it ought to come into being" as St. Augustine declares. Such is the authentic evolution as recognized in Patristic tradition, which constitutes not the creation of something new, but indeed an "unfolding," as the very term implies, of something which already exists. At a certain moment, "when it ought to come into being," that organism breaks through, as it were, into space-time and thus becomes perceptible; and this "breaking through" constitutes its second birth, its birth in the ordinary sense.[6]

It is to be noted that this conception of a "double birth" is in fact biblical. It is briefly but yet unequivocally stated in Chapter 2 of Genesis, verses 4 and 5, which deviate abruptly from the perspective of the hexaemeron, as described in Chapter 1. In that remarkable passage, which seems to have been rarely understood, the first birth is relegated to *"the day when the Lord God made the heaven and the earth,"* whereas the second is depicted as an emergence into the space "above ground," as of a plant springing up in the earth. The more deeply one probes these two crucial verses, the more clearly one perceives that they provide the key to the mystery of biological origins.[7]

Such, in brief, is the biblical doctrine concerning biogenesis. Unlike "creationism" of the customary kind, it opens ontological vistas unsurmised by the opponents no less than the protagonists of Darwinism. From an authentically ontological point of vantage one sees at a glance that Darwinian transformism constitutes but a brutal attempt to explain the mystery of origins on the corporeal plane, where precisely it can *not* be resolved. We know today in light of quantum theory that not even an electron fits into that truly narrow world—what to speak of plants and animals, and above all, of man himself.

6. For a summary of this Patristic doctrine I refer to pp. 20–25 in my book, *Theistic Evolution: The Teilhardian Heresy* (Tacoma, WA: Angelico Press/Sophia Perennis, 2012). Extensive documentation may be found in Fr. Seraphim Rose's monumental treatise, *Genesis, Creation, and Early Man* (Platina, CA: St. Herman of Alaska Brotherhood, 2000).

7. For an exegesis of the verses in question I refer to *Theistic Evolution*, op. cit., pp. 249–252.

4

Celestial Corporeality

Leiblichkeit ist das Ende der Werke Gottes
Friedrich Christoph Oetinger

Having reflected at length upon the nature of the corporeal plane and touched upon the intermediary, it behooves us now to consider the third and highest cosmic domain, which is the spiritual or celestial, as it is also called. I propose, moreover, to view that ontological domain in its lowest aspect or modality, which may appropriately be termed *celestial corporeality*.

As is well known, Christianity believes not only in the immortality of the human soul, but also in what it terms the resurrection of the body. Somehow the bodies of the deceased will be "raised," and in the process, transformed and glorified. "*Behold*," St. Paul declares, "*I show you a mystery: We shall not all sleep, but we shall all be changed. In a moment, in the twinkling of an eye, at the last trump: for the trumpet shall sound, and the dead shall be raised incorruptible, and we shall be changed. For this corruption must put on incorruption, and this mortal must put on immortality.*"[1]

Whatever may be the position of other religions in that regard, Christianity does not place before us the prospect of a discarnate posthumous state: "*For we that are in this tabernacle do groan, being burdened: not for that we would be unclothed, but clothed upon, that mortality might be swallowed up of life.*"[2]

This miracle, this unthinkable prodigy, the Christian maintains, has already come to pass in the resurrection of Christ, "*the firstborn*

1. 1 Cor. 15:51–53.
2. 2 Cor. 5:4.

from the dead."[3] The dogma of bodily resurrection, moreover, is absolutely crucial to Christianity: *"If there is no resurrection of the dead,"* St. Paul affirms, *"then Christ is not risen. . . . And if Christ be not raised, your faith is vain."*[4]

Admittedly that teaching has always been "hard," and offensive to the philosophers. It is easy enough to imagine or to conceive of a discarnate state; but the raising of dead bodies—that is something else. No wonder the learned men of Athens shook their heads in disbelief. *"And when they heard of the resurrection of the dead,"* we are told, *"some mocked: and others said, We will hear thee again on this matter. So Paul departed from among them."*[5] Nor did the skepticism and disdain of the intellectuals abate with the spreading of the faith. One is reminded of the Platonist Celsus who so eloquently berated Christianity: "a religion befitting an earthworm" he called it. What is new, in our day, is that theologians professing to be Catholic have joined the skeptics, and have felt obliged to expunge the offending dogma by one means or another. In so doing, however, they have contradicted the explicit and indeed *de fide* teaching of the Catholic Church, and have in fact struck at the very heart of Christianity. Once again: *"If there is no resurrection of the dead, then Christ is not risen. . . . And if Christ be not raised, your faith is vain."* For the Christian, thus, the stark spectacle of the "empty tomb" is not negotiable. It cannot, as some contemporary *periti* would have us believe, be written off as a literary device intended to bring home the idea of immortality to unphilosophic minds. Christianity maintains, on the contrary, that the "empty tomb" testifies to the greatest miracle ever witnessed on earth: the bodily resurrection of Jesus. Christians believe, moreover, that a similar resurrection will take place universally "at the end of time," when Christ will come again: *"in the clouds of heaven, with power and great glory."*[6]

Now, the pressing problem for theology—and indeed for metaphysics—is to elucidate the dogma of bodily resurrection, to render

---

3. Col. 1:18.
4. 1 Cor. 15:13, 17.
5. Acts 17:32, 33.
6. Matt. 24:30.

it somehow conceivable. What Christ and His Church have taught concerning the nature of God, or of the human soul, is by comparison far less difficult it would seem, far less problematic to the human intellect. The hardest dogmas, it appears, are those that hinge upon the notion of *body*, the idea of "flesh and blood" if you will, transposed into the spiritual realm. Think of the well-nigh unbelievable words of Christ when He taught in the synagogue at Capernaum: "*Verily, verily, I say unto you, Except ye eat the flesh of the Son of man, and drink his blood, ye have no life in you.*"[7] The Jews moreover understood well enough that this was more than a mere figure of speech, a mere metaphor: "*Many therefore of his disciples, when they heard this, said, This is an hard saying; who can hear it?*" Meanwhile the Catholic Church continues to insist that the *flesh* and *blood* in question are to be taken literally—with the understanding, of course, that these terms refer precisely to the risen and glorified body of Christ. It is this glorified or "celestial" body that is offered as food in the sacrament of the Eucharist. But let us not forget that the host, when consecrated, does not disappear; it is not simply replaced by the body of Christ, but is transfigured into that body: it is "transubstantiated" to use the official term. What takes place, thus, upon every Catholic altar is akin to the Resurrection: it is one and the same Christic miracle that presents itself in two modes.

The mystery of celestial corporeality and its attainment—be it in the Resurrection or in the Mass—has of course been speculated upon by theologians since apostolic times. Yet it appears that some sixteen centuries into the debate a Christian layman—a cobbler and peddler of clothing, no less—was able to contribute decisive insights, which perhaps to this day have been insufficiently reflected upon. I am referring of course to Jacob Boehme, whose teachings can be viewed as inaugurating a new school of theological speculation. In the present chapter I propose, first of all, to give a brief overview of Christian thought as it bears upon the nature of celestial corporeality, from its apostolic beginnings up to the end of the sixteenth century. These historical observations, moreover, are based upon a paper by Julius Hamberger, which appeared in the

7. John 6:53.

Jahrbuch für Deutsche Theologie in 1862.[8] I might add that Hamberger was a disciple of Franz von Baader, the renowned Catholic exponent of Boehme's thought, and was himself a profound thinker who deserves to be far better known than has so far been the case. Having completed the aforesaid historical overview I propose to present the relevant conceptions of Jacob Boehme, following closely the interpretation of Pierre Deghaye.[9] What I hope to convey is the fact that Boehme's doctrine enables us to view the subject of "*himmlische Leiblichkeit*" in a distinctly new light. Following this, I propose to reflect further upon the nature of celestial corporeality, in an effort to distinguish the celestial as clearly as possible from the kind of corporeality known to us here below. The final section will focus upon "time and eternity," and it is here that the preceding considerations prove pivotal.

Hamberger begins his historical essay with a summary consideration of "*himmlische Leiblichkeit*" as conceived in various pre-Christian traditions. However, considering the extreme difficulty of penetrating, *par distance* and from the outside, teachings such as those of ancient Egypt or of Kashmiri Tantrism, let us say, it may be best to leave out of account the opinions of the German savant regarding such domains. The case is different when it comes to the teachings of Plato and Aristotle, not only because they stand at the beginning of what may be termed the philosophic mode, but also because both have exercised a profound and indeed decisive influence upon the development of Christian thought. It will therefore be appropriate to begin our recapitulation of Hamberger's essay at that point.

He contends, first of all, that the doctrines of both Plato and

8. "Andeutungen zur Geschichte und Kritik des Begriffes der himmlischen Leiblichkeit," vol. 7, pp. 107–165. I am indebted to Professor Roland Pietsch of the Maximilian University at Munich for bringing this material to my attention.

9. One of the foremost authorities in this field, Deghaye has provided what may well be the finest commentary on the thought of the German mystic. See especially *La Naissance de Dieu ou la Doctrine de Jacob Boehme* (Paris: Albin Michel, 1985).

Aristotle—each in its own way—rigorously exclude what he deems to be the authentic Christian conception of celestial corporeality. "Plato no less than Aristotle," he writes, "despite the fact that they recognized a supreme unity, transcending every opposition, as the source of all being, remained nonetheless confined within a dualism of the ideal and the real, of spirit and matter." In either philosophy a permanent and perfect union of spirit and matter, of soul and body, Hamberger believes, is unthinkable. One is forced to conclude in either case that the opposition between the real and the ideal, the sensible and the intelligible, is transcended nowhere save in the supra-ontological sphere of the Absolute. An elevation of the material to the spiritual plane is unthinkable; St. Paul was right: that Christian tenet *is* "foolishness to the Greeks."

The earliest indications, in Judaic tradition, bearing upon celestial corporeality are found in the accounts of Enoch and of the prophet Elijah. What we are told in the case of Enoch is exceedingly sparse: when he was three hundred and sixty-five years old, we read in Genesis 5, "Enoch walked with God; and he *was* not; for God took him." Having lived an exceptionally pious and virtuous life, "God took him," and he was seen no more. As Hamberger observes: "The body of this patriarch, one may gather, was delivered from earthly existence by being swallowed up, as it were, by the life of the spirit, or better said, taken up into that life and thus brought to transfiguration." A similar event is recounted in the case of Elijah: one day, as the prophet was walking with Elisha, his son, "it came to pass, as they still went on, and talked, that, behold, there appeared a chariot of fire, and horses of fire, and parted them both asunder; and Elijah went up by a whirlwind into heaven." (II Kings 2:11) The graphic and explicit nature of this account suggests that it may refer to an actual experience of a visual kind; and yet, as Hamberger points out: "No pious Israelite could have doubted that, behind what the bodily senses could perceive, an event has taken place that extends into the realm of the invisible, the transformation, namely, of the material corporeality of the prophet into the supra-material, even if that Israelite was unable to understand clearly the essential difference between the two."

The definitive texts, however, on the question of *himmlische*

Leiblichkeit are evidently those of the New Testament relating to the resurrection of Christ. It is here, in this unique and epochal event, that celestial corporeality enters, as it were, into our world: enters upon the stage of history, one might say. What happened on that first Easter Sunday is indeed, in a sense, historical; speaking of the risen Christ, St. Paul assures us that "*he was seen of above five hundred brethren at once.*" It is easy, however, to misinterpret the biblically attested facts, and thereby miss the very point of the Resurrection. One needs to realize that this ostensibly historical event marks nonetheless a radical break with the earthly condition: "From that moment on," Hamberger observes, "the Savior belongs no longer to the earthly realm, but to a higher sphere, from whence he reveals himself to his disciples only in sporadic appearances, as if by visitation." The Savior, one may presume, revealed himself to his yet earth-bound disciples in visible and tangible form to convey to them the fact that he has indeed risen from the dead; and yet, when he appeared suddenly behind locked doors or ascended miraculously, he likewise demonstrated that his spirit had now attained total dominion over his corporeal nature. "Thus his corporeality," writes Hamberger, "was now evidently supra-material and no longer subject to the bounds of time and space."

All the same, many among the followers of Christ have conceived of his transfigured corporeality in basically material terms, suitably "thinned or sublimated" as Hamberger notes. This widespread and indeed dominant tendency has moreover been complemented by an opposing trend, epitomized by the rigorous spiritualism enunciated by Origen. As Hamberger explains: "The reason for the spiritualistic course which Origen pursued lay no doubt in his concern to purify the conception of spirit from the contamination of every material admixture, which however seemed to him to necessitate the exclusion of all corporeality, because, with Plato, he regarded the perfect reconciliation of spirit and nature, the complete elevation of the second to the first, as impossible. But inasmuch as this spiritualism proved to be incompatible with the true sense of Scripture, there arose against him determined opponents, who sought with great emphasis to defend the biblical realism, yet could not do justice to the biblical truth."

The foremost among these "determined opponents" was no doubt St. Jerome, who championed what might nowadays be termed the fundamentalist position. True flesh, for him, was only "what consists of blood, veins, bones, nerves and the like; also teeth, stomach, and genitals could not be absent in the celestial state." It may however come as a surprise that even St. Augustine, Christian Platonist though he may have been, should have held similar views on the subject, "yet in a more delicate form" as Hamberger observes. The reason, however, for this apparent incongruity, is not far to seek: "Without doubt, Augustine was entirely right when he refused to view the transfiguration of the body as its cancellation or disappearance; the true character, however, of the transfigured body, which consists in a perfect harmony and concord of nature with spirit, of the real and the ideal, he failed to recognize. Against the former error he was protected by his faithful adherence to the word of Scripture; the latter recognition, on the other hand, may have eluded him on account of his predilection for Platonic philosophy."

Be that as it may, it does appear that Christian speculation regarding celestial corporeality has tended to vacillate somewhat between two extremes: the fundamentalist and the spiritualistic, one might say. But as Hamberger points out: "Besides such erroneous deviations to one side or the other, there have never been lacking in the Church teachers who knew how to hold fast to the biblical sense of celestial corporeality, and employ that conception in the unfolding of systematic theology." An excellent example would be Tertullian. That this Patristic author was far removed from the spiritualistic camp of Origen is obvious, given his famous statement: "Who will deny that God is a body, even though He is a spirit." Yet the very boldness of this assertion makes it clear that Tertullian is also worlds removed from the fundamentalism of St. Jerome. "Nothing was further from Tertullian's mind," writes Hamberger, "when he ascribed corporeality to God, than to conceive of that corporeality after the fashion of our earthly body." And it is of interest to note that St. Augustine, while finding it hard to penetrate Tertullian's language, regarded him nonetheless as basically orthodox. Tertullian "did not wish to depart," Hamberger writes, "from what the Bible itself suggests quite clearly concerning the corporeality of

God, and also recognized well enough the truth that all reality must be somehow *formed*, and that, in the absence of some kind of corporeality, the spirit could not function as such." For Tertullian spirit and body are correlatives: the one demands the other. "All invisible things," he writes, "have for God their body and their form by which they are visible to Him." As Hamberger observes: "One sees clearly that Tertullian was permeated by the thought of supra-material corporeality." It is likewise evident, however, that this predilection for supra-material corporeality has never been shared by the vast majority of theologians.

Yet it happens that the Christological teachings enunciated, two centuries later, at the Council of Chalcedon are in fact concordant with Tertullian's idea of supra-material embodiment. Hamberger has it exactly right: "When namely it is ascertained at Chalcedon that one cannot, on the one hand, regard the human nature of Christ as having been absorbed by the divine, but neither can one regard it as existing in opposition to the divine, or even, as it were, side by side, but must instead assume that the human nature is entirely permeated by the divine and received into the same, it follows that something similar must then apply in general to celestial corporeality in relation to the life of the spirit. In consequence of the aforesaid conciliar decision, one is not permitted to think that in this elevated state the flesh has disappeared, as if absorbed by the spirit; but neither may one conceive of the transfigured flesh in material terms." It appears, however, that the teaching of Chalcedon was generally viewed exclusively in its Christological aspect, and that its implications regarding celestial corporeality as such have generally been ignored. Thus St. John Damascene, to cite just one example, does not hesitate to speak of "deification" with reference to the glorified flesh of Christ, but continues all the same to conceive of the celestial bodies of angels and of saints as inherently material entities.

To be sure, there have also been theologians who did perceive the wider implications of the Chalcedonian position. This holds true, above all, in the case of John Scotus Eriugena, for whom the idea of celestial corporeality occupies a position of central dominance. As Hamberger observes in that regard: "Not only does this very conception appear in its highest purity, but also in such a universal manner as to pervade the entire doctrinal system of this great thinker and render possible an authentically scientific understanding of Christian truth in its main elements." What John Scotus explicates with utmost clarity and emphasis is the supra-materiality of the first creation: "God is immortal," he declares, "and what He alone creates, that is likewise immortal." In keeping with this idea, he conceives of the original state of man in distinctly celestial terms. He does not hesitate, in fact, to apply the dogmatic formulas of Chalcedon to mankind, and indeed, to the universe at large, in their state of perfection. His doctrine, as it applies to the perfected universe, constitutes in effect a Chalcedonian cosmology. John Scotus is able to conceive of a universal transfiguration precisely because he views the matter of our world, not as something absolutely primordial, but as itself derived from an immaterial principle. "There is nothing in human nature," he declares, "that is not spiritual and intelligible; even the substance of the body is intelligible." Materiality makes its appearance only where spirit or will stand in opposition to the archetypal idea, that is to say, only where "sin" has entered. "It is unthinkable," says John Scotus, "that the body was perishable and material before the cause of death and materiality, namely sin, had appeared." This position, however, has eschatological implications; as Hamberger points out: "The re-elevation to supra-materiality becomes thus possible through the conquest of sin, and the way is thereby opened for a return of the world to God, to the end that God may become all in all."

John Scotus conceives the supra-material state to be supra-spatial and supra-temporal as well. As surely as God Himself is above time and space, he maintains, it is certain that, with the elevation of the presently material world, "time, as the measure of motion, will disappear, and so too space, as the separator of things, shall be no more." Both space and time are aberrant, moreover, precisely because, in the celestial order, there *are* no separations of that kind. "There is nothing incredible or irrational," writes John Scotus, "in the supposition that intelligible beings unite, so that they are one, while each retains its own characteristics, but in such a way that the lower is contained in the higher." To help us understand, he gives the example of air permeated by the light of the Sun, which though it pervades the air yet retains its own substance. "Likewise, I believe, will the corporeal substance go over into the soul, not to perish, but that, having been elevated to a more excellent condition, it shall be preserved. So too one must suppose that the soul, having been received into the intellect, becomes more beautiful and more similar to God. In the same way I think of the entry into God— not perhaps of all, but certainly of rational substances—in whom they shall reach their goal, and in whom they shall all become one."

This grand and majestic vision of God, man, and universe, however, seems never to have been shared by many—not, in any case, where corporeal nature is concerned. In Hamberger's words: "As little as Tertullian was able to gain for his profound spiritual intuitions a wide recognition, so too John Scotus would not be able to convey his sublime worldview to his own and future generations as their common heritage." Whatever the reasons for this resistance or neglect may be, John Scotus' lofty conception of celestial corporeality has apparently had little impact upon the subsequent course of mainstream theological speculation.

The case of Albertus Magnus and his illustrious disciple is of course of special interest. Now, it does appear that Albertus may have been open to John Scotus' idea of celestial corporeality, although it remains unclear whether he was in any way influenced by the Irish theologian. It seems more likely that Albertus derived inspiration from Avicebron, whose writings he studied assiduously, and whose thought was in certain respects kindred to the teachings of John Scotus. As Hamberger explains: "Avicebron namely was sharply opposed to the usual assumption that substance composed of matter and form must always and everywhere be thought of as something material. The concept of the material, he taught, rests upon the notion of quantity, whereas substance need not entail quantity, but may well be conceived without quantity, and hence without materiality, while still composed of matter and form." But whereas this position—which apparently Albertus did not oppose—permits the conception of a supra-material corporeality, the German master seems not to have fully availed himself of this option. According to Hamberger, Albertus Magnus attained to the true concept of *himmlische Leiblichkeit* "only in the case of the God-man, and only with difficulty can his utterances concerning the bodies of the risen saints be reconciled with the idea of an actual transfiguration." When it comes to St. Thomas, moreover, Hamberger thinks that on this issue Aquinas reverted more or less to the position of Jerome and Augustine. In support of this contention he points out, for example, that the blessed, according to the teaching of St. Thomas, "shall appear at their resurrection in the age of their youth, namely, in that period of life which stands midway between growth and decline"—a view which admittedly is worlds removed from the thought of John Scotus. Hamberger maintains, however, that this "insufficient, and indeed erroneous conception of the resurrected body" is to be ascribed, not to any deficiency on the part of St. Thomas himself, but to external constraint: "It was not an inner, but an outer fetter that impeded the flight of his thoughts." In any case, the fact remains that the theological direction championed by Tertullian and John Scotus did not flourish—for whatever reason—within the bounds of the official Roman Catholic Church. The further development of this theological direction is rather to be found near the periphery, so to speak, in the aftermath of the Reformation, and chiefly in Protestant lands. One must look for it, not among theologians of ecclesiastic rank, but in the circles of mystics, physicians, and God-fearing philosophers. It was moreover in the teachings of our pious and lowly cobbler that this movement, if we may call it such, reached what appears to be its fullest expression, and possibly its consummate form.

German theosophy did not originate abruptly with Jacob Boehme: the movement has earlier representatives, the most illustrious being Paracelsus (1493–1541) and Valentin Weigel (1533–1588). Even so, Boehme's thought displays the originality and titanic force of a revelation; and as is typically the case with a legacy of this kind, his teaching proves to be *de facto* inexhaustible. After centuries of analysis and commentary the fascination of this enigmatic figure remains undiminished. Today, in fact, standing at a greater distance, we can all the better perceive the seminal potency of his thought. Many literary strands and esoteric traditions have met and mingled in Boehme's system; but in the process they have become radically transformed: what emerges is a coherent teaching bearing the distinctive signature of the *philosophus teutonicus*. At his hands our perceptions of God, man, and the universe—while remaining rigorously Christian and indeed biblical!—have been profoundly altered and ostensibly enlarged in certain respects. If St. Thomas Aquinas has enriched Christianity by "christianizing" the wisdom of Aristotle, it can indeed be said that Jacob Boehme has done likewise with reference to the Hermetic tradition. In his opus a profound and long-forgotten "alchemical" philosophy of nature reveals its contours, but in a new context, a brand new key; as Pierre Deghaye explains: "Transposing philosophy of nature to the level of a supreme knowledge proper to theology, Boehme made of it a *theosophy*."[10] And he goes on to say: "In fact it is also a theology, that is, a science of God. But it is profoundly different from dogmatic theology, not only Lutheran,[11] but of any confession whatever. Theosophy represents another approach to God. . . ."

The genre itself did not originate with Boehme; for as Deghaye points out, Jewish Kabbalah and Islamic mysticism are likewise theosophies. "What these three types of thought have in common," he writes, "is that their subject matter is God making Himself known." Boehme, in particular, distinguishes clearly between God as Absolute—what he signifies by the term *Ungrund*—and God as He is revealed, not only to humanity, but first of all, to Himself. There can be no science, he maintains, no knowledge of any kind, of

10. Our quotations are taken from "Jacob Boehme and his Followers," *Modern Esoteric Spirituality*, Antoine Faivre and Jacob Needleman, eds. (New York: Crossroad, 1995).

11. Though born in Lutheran territory and nominally Lutheran, Boehme was in fact persecuted by the Lutherans, to the point that he could not have even be given Christian burial if it had not been for the intervention of an influential friend. But whereas Boehme's doctrine is indeed at odds with Luther's ideas, it is by no means antagonistic to Catholic theology. Allowing for the fact that theosophy and dogmatic theology represent different points of view, it can be argued that Boehme stands yet within the integral Catholic tradition, which is in fact *katholikos*.

the *Ungrund* as such: only the self-manifesting God can be the object of knowledge, and indeed, of love. Basing himself, one has reason to believe, upon a mystical experience, Boehme envisages a process of self-revelation *in divinis* which, theosophically speaking, "gives birth to God." It is this "eternal birth," moreover, that constitutes the primary theme of Boehme's theosophy. Supra-temporal though it be, Boehme conceives of that "birth" in terms of a seven-fold cycle, which in fact constitutes the archetype of time. As Deghaye observes, it represents "the week of creation transposed to the level of an utterly first origin." Speaking in temporal terms, the cycle can be described as a conquest, in successive stages, of a primordial darkness or chaos, which Boehme also refers to as a dark fire. "Dogmatic theologies," Deghaye points out, "speak first of the light that is synonymous with divine perfection. They mention darkness only in reference to the angel precipitated into it. Boehme puts darkness first. The first part of the seven-part cycle of divine manifestation is dark. In order for light to pour forth, it must shatter the darkness." God's birth constitutes thus a Victory, a supreme Heroic Act which in fact prefigures the Death and Resurrection of Christ. Strange as it sounds, it can indeed be said that "Boehme's God dies before he is born." The defining event of Christianity comes thus to be viewed in a strikingly new perspective: "What transpires on earth once Christ has come among us," writes Deghaye, "only objectivizes for us the primordial event that unfolds in the seven-part cycle."

Now God is revealed, is rendered visible, in His glory, which is a radiant body, a body made of celestial light. We must not however think of the revealed God as existing simply by Himself, "in splendid isolation"; as Deghaye explains: "In order to contemplate himself, God requires that a mirror be offered to him in a form he has raised up, which, although inhabited by him, is distinct from him. . . . The mirror is the body of the angels. The revealed God appears the moment angels are there to contemplate him." One sees that the birth of God is accompanied by the genesis of His eternal abode, which is the primordial heaven, the celestial realm. Boehme's doctrine, thus, is not only a theogony, but a cosmogony as well. It constitutes in fact a triple cosmogony, which is to say that

Boehme envisages the genesis of *three* worlds, corresponding to what he terms the three principles. The highest of these is the primordial heaven, which is "a world of light." But this celestial realm, Boehme maintains, is brought into being by way of a sevenfold cycle that begins in darkness: a darkness which, strange to say, already prefigures Hell. It is the fall of Lucifer that objectivizes this primordial darkness, and in so doing, gives birth to the infernal realm. Thus, in wanting to be "like unto God," Lucifer has inverted the very process that founds the angelic domain: instead of light "shattering the darkness," it was now the darkness that shattered the light of his angelic nature; and so "a world of darkness" came to be. Our world, finally, is the third; it is preceded by the creation of Adam and precipitated by his Fall. A temporary world according to theosophy, its *raison d'être* is to permit the reconciliation of a fallen humanity with God. By its very nature it constitutes an intermediary realm in which Light and Darkness—the heavenly and the infernal principles—are mingled. Boehme refers to it as the third principle. One might add that this intermediary status, which theosophy ascribes to our world, is indeed borne out by its phenomenology; for as we all know well enough, ours is a world of combat, of ceaseless strife impelled by the clash of opposing forces.

What proves to be of major interest in Boehme's theosophy is the relation of our world to the heavenly, the primordial, which is its archetype. "The first nature," writes Deghaye, "engenders the other nature, our own, which obscures it while at the same time manifesting it sufficiently to reflect it. This second nature will be destroyed [at the end of time] and the primordial nature unveiled, manifesting God in all his Glory." Despite the dissimilarity of the two—which in fact puts the first beyond even our imaginative reach[12]—the connection between them is exceedingly close; as Deghaye has put it: "These two degrees appear successively, but the two natures coexist. Eternal nature remains within the envelope of our nature . . . and yet there cannot be the least confusion between the two." The distinction, however, between "inner kernel" and "outer shell"—the

---

12. I am leaving out of account what Henri Corbin refers to as the active imagination, which constitutes an inherently spiritual faculty.

fact that there is not the least confusion between the two—must not be misconstrued as an ordinary separation; one must not forget that "the two natures coexist," and that the second in fact depends upon the first. The "inner kernel," thus, so far from constituting a kind of foreign body, is to be conceived rather as the "kernel of reality," of which the "shell" is but the outer manifestation.

It is becoming apparent that Boehme's doctrine is intimately related to the alchemical quest, and presumably validates whatever truths that age-old tradition may enshrine.[13] One is tempted to surmise, moreover, that Boehme's conception of a primordial and exemplary cycle may indeed harbor the rudiments of a universal science, applicable in principle to every natural domain; and it is not without interest to note that at least one contemporary scientist has drawn inspiration from Boehme's writings in the pursuit of his own discipline, which is that of particle physics.[14] What the Master himself has his eye upon, however, is indeed nothing less than the salvation of man, what he terms our second birth. That is the one and only "alchemical transformation" the German mystic pursues relentlessly; and he does so as a believing and pious Christian. "Our second birth," writes Deghaye by way of commentary, "is the equivalent of the resurrection of Christ, and is anticipated in the cycle of primordial nature. What takes place in this exemplary transforma-

13. Whereas the educated public at large remains no doubt skeptical, if not indeed disdainful of alchemy, there is today a growing awareness among the better informed that what stands behind many an "exploded superstition" may in fact be a long-forgotten truth. Take the case of the so-called four elements: earth, water, air and fire. One can be reasonably certain that these terms were not employed alchemically in their ordinary sense, but were used to designate *elements*, precisely, out of which corporeal substances are formed. Somewhat like the quarks of modern physics, these elements are not found in isolation, but occur in their perceptible combinations. Now, given the fact that corporeal substances are not in truth "made of quantum particles" (as I have argued repeatedly, beginning with *The Quantum Enigma*, op. cit.), it is by no means unreasonable to suppose that there may indeed be such alchemical elements. It turns out that our habitual opposition to alchemy is based mainly upon a reductionist prejudice, for which there is in reality no scientific support at all.

14. Basarab Nicolescu, *Science, Meaning, and Evolution: The Cosmology of Jacob Boehme* (New York: Parabola Books, 1991). See also my review in *Sophia*, vol. 3, no. 1 (1997), pp. 172–179.

tion at the threshold of time is already a death and a resurrection. The cycle of origins is simply repeated each time life blossoms forth following transformation. All life is born only to die and be born a second time. The theosophy of Boehme is a theology of the second birth. This is what connects it with Christian mysticism, in which the focal subject is the birth of Christ in us."

Now, all birth is an embodiment, an incarnation, a certain union of spirit with flesh. Not all flesh, however, is of the same kind. St. Paul speaks of this clearly in the fifteenth chapter of First Corinthians, in the celebrated discourse in which he distinguishes between corruptible and incorruptible bodies, the earthly and the heavenly. Our first body is of the earth, earthly; our second shall be celestial, heavenly (*ex ouranou*). If Boehme's doctrine, therefore, is indeed a "theology of the second birth," it is also, and by the same token, a theology of celestial corporeality. There is in fact reason to surmise that the concept of *himmlische Leiblichkeit* comes into its own in the theosophy of Jacob Boehme.

All bodies, to be sure—be they ever so celestial!—are made of a substance of some kind; and when it comes to celestial bodies, that substance, as one would expect, is indeed the most precious, the most refined, the most excellent of all. "The eternal heaven," writes Deghaye, "of which the bodies of angels are made, is the precious material that develops over the course of a seven-part cycle described as the divine masterpiece." It is best conceived as a pure radiance, a celestial light, of which what we know as light here below is but a pale reflection. This celestial light, moreover, is none other than God's glory, the very radiance that constitutes what theosophy terms "the body of God." That "Body," therefore, no less than the bodies of angels and of the saints in heaven, is composed of the same "precious material" born of Primordial Nature's dark and fiery womb: "Light is its substance, its flesh" as Deghaye declares.

We are now, finally, in a position to appreciate the words of Friedrich Christoph Oetinger which I have placed at the head of this chapter: "Corporeality is the end of the works of God." The consummation of God's work of creation is indeed *embodiment*: a union of spirit and flesh, in which the two become as one. It is that perfect submission, that complete "transparence" to the indwelling spirit, that is realized in celestial corporeality. As Deghaye profoundly observes: "The only true perfection is that which is incarnated in a body of light."

"The first nature," it has been said, "engenders the other nature, our own, which obscures the former while at the same time manifesting that first nature sufficiently so as to reflect it." That obscuration, however, is such that we are unable—even in our wildest flights of imagination!—to picture, or somehow present to ourselves, that radiant and eternal nature which underlies our own. The two natures—the two worlds!—actually coexist, which is to say that the first penetrates the second, not spatially, to be sure, but ontologically. In a word, it lies "within," as Christ Himself testifies: "*Neither shall they say, Lo here! or lo there! for, behold, the kingdom of God is within you.*"[15] That *regnum* or kingdom, moreover, proves to be none other, theosophically speaking, than the first and celestial nature which lies concealed here below, not only in our human flesh, but indeed in all things. Truly: "*The light shineth in darkness, and the darkness comprehended it not.*"[16] Boehme may indeed have been the first to penetrate the cosmological implications of this Johannine dictum, which in fact epitomizes his own conception of our world as the so-called "third principle."

Despite its ontological proximity, thus, celestial nature remains for us literally shrouded in darkness. Yet we incline to picture celestial realities in basically earthly terms: even the most eminent theologians, as we have had occasion to observe, have often enough, apparently, succumbed to this human-all-too-human tendency. To do so, however, is not only to falsify, but in a sense to invert the truth; as Hamberger explains:

> It would be totally wrong (*völlig verkehrt*), when conceiving of celestial corporeality, to retain in some way the idea of earthly materiality, in order not to forfeit the nature of corporeality itself. Be it that one imagines that higher corporeality as a perfected

15. Luke 17:21.
16. John 1:5.

earthly substance, elevated to the highest, noblest form, or that one thinks of it as drastically rarefied—neither in the one nor in the other case would one have reached the true conception. The celestial stands simply above the earthly, and every imperfection to which the earthly remains subject, even when sublimated to the highest degree, must be altogether excluded from the celestial. Every earthly admixture, even the faintest breath, that stems from this nether world, would contaminate the thought of celestial nature, and indeed, would straightway cancel the same.[17]

In the terrestrial domain corporeal entities "occupy space," which is to say, admit extension; we need thus to ask ourselves whether the same applies in the celestial world. Now, it is clear from the start that "celestial space"—if indeed there be such a thing—does not simply coincide with the space we know by way of sense perception. Celestial bodies are not situated, properly speaking, in *our* space: "*Neither shall they say, Lo here! or, lo there! for behold, the kingdom of God is within you.*"[18]

But by the same token one cannot say, of a celestial entity, that it is *not* here, or *not* there; for this too would, in a way, subject the being in question to the conditions of terrestrial space. So too, when one speaks of the celestial realm as residing "within," it needs evidently to be understood that this too is not meant in a literal, that is to say, a *spatial* sense. But whereas celestial bodies remain thus aloof from any and all spatial conditions pertaining to *our* world, it cannot be denied that there must be something in the celestial realm that corresponds to the idea of space, and in fact constitutes its archetype.

On this vital question Julius Hamberger has something exceedingly important to contribute. Here is what he writes:

17. Our quotations from Julius Hamberger in this section are taken from a second paper, likewise published in *Jahrbuch für Deutsche Theologie*. It is entitled "Die Rationalität des Begriffes der himmlischen Leiblichkeit," and appears in Vol. 8 (1863), pp. 433–476. After contributing three major articles on *himmlische Leiblichkeit* to the *Jahrbuch* between 1862 and 1867, Hamberger published this material in book form, under the title *Physica Sacra, oder der Begriff der himmlischen Leiblichkeit und die aus ihm sich ergebenden Aufschlüsse über die Geheimnisse des Christenthums*. The book appeared in Stuttgart, in 1869.

18. Luke 17:21.

Not infrequently one is reluctant to attribute extension to celestial formations for fear of degrading them thereby to the level of materiality. Admittedly, an extension such as one finds in earthly things cannot apply to them; however, what is in no sense extended could have no being, no reality at all. Even spirit must have a certain extension; one must not let it be shrunk into a narrow confine, or better said, into the nothingness of a mathematical point, if it is actually to exist. When however earthly formations, on account of the prevailing obstructions affecting their life, move apart into the breadth of terrestrial space, the same cannot apply to the fully vital beings of the celestial realm. Since these admit of no inner separation, their extension must indeed be of an intensive kind, which is to say that they do not extend outwards, but inwards, not in breadth, but in depth; and this depth stands as much above that breadth as eternity looms above time.[19]

The notion of an "intensive extension" is of course far from clear; and yet it is highly suggestive. One begins to sense that there *must* in fact be a state answering to a conception of this kind. The idea is moreover reminiscent of the "inner space," the "space within the lotus of the heart" described in the *Chandogya Upanishad*, which is said to contain "both heaven and earth, both fire and air, both the Sun and the Moon, lightning and the stars." In short, all things whatsoever are to be found in that inner space, which from our mundane point of view has no extension at all. They are, however, to be found in that interior space, not in their outward manifestation, but in their innermost being, precisely, which coincides, as we have seen earlier, with their celestial archetype. The "space within the lotus of the heart" can thus in truth be none other than celestial space, and if it contains all things, it is by virtue of the fact that all things are indeed celestial at their core.

One sees that the inner space is by no means like the outer; and if indeed that external space constitutes an image of the celestial,

---

19. As another Boehme friend observed: "In eternity one thing does not stand outside another, as in our relation of material space, but the one thing is in the other, and yet is different from it." (H.L. Martensen, *Jakob Boehme: His Life and Teaching* [London: 1885], p. 74.) The affinity in all this to the conceptions of John Scotus Eriugena is very much in evidence.

Hamberger's notion of intensive extension suggests that the image is actually *inverted*. There is not only an analogy, therefore, but an opposition as well between the two; and it therefore requires a radical reorientation—an authentic *metanoia* in the ancient sense—on the part of the percipient to pass from one kind of spatial perception to the other. And this, to be sure, must be the reason why we find it so difficult—nay, impossible in our present state—to grasp the nature of celestial bodies; as one reads in the *Bhagavad Gita*: "In that which is night to all beings, the in-gathered man is awake; and where all beings are awake, there is night for the *muni* who sees." No use trying to picture to oneself what "the in-gathered man"—the *samyamî*—beholds; *his* world is indeed "night" to us. The ontological fact is that spatial separation, or distance as we know it, has no place in the celestial realm, in which extension occurs "in depth."

The external space, one can say, separates, whereas the inner unites; the former is by nature centrifugal, while the latter is centered upon God.[20]

An Aristotelian definition will help us to understand this difference more clearly. Quantity, says Aristotle, is that which "admits mutually external parts." Take the case of a line segment: not only does a segment, when bisected, break into two mutually external pieces, but these pieces are truly parts. The point is that the whole, in this instance, is indeed the sum of its parts—and not something more, something which is therefore inherently indivisible. Similar considerations apply to number, that is to say, to discrete as opposed to extended quantity. Even the sophisticated mathematics of our day, in all of its numerous branches, stands yet under the aegis of quantity in the precise Aristotelian sense. And this is only as it should be: *mathematics is indeed the science of quantity.* Not everything, however, is quantitative. Take color, for example: red or green, obviously, does not admit of mutually external parts. Red

---

20. This inversion has been strikingly portrayed by Dante in the *Divina Comedia,* in which the terrestrial and celestial realms are poetically combined into what mathematicians nowadays would term a 3-sphere. In this "model" the two realms—the celestial and the terrestrial—are viewed as the northern and southern hemispheres, respectively, of that 3-dimensional sphere, separated by the Empyrean, which constitutes the equatorial 2-sphere. See Chapter 8, pp. 169–172.

bodies, of course, are divisible; but redness as such is not. This means, in particular, that one does *not* change the color of an object by breaking it into pieces: color comes along "for the ride" if you will, without being in any way affected by division. Color, therefore, is not a quantity, but a *quality* as we say, and thus a thing which cannot be comprehended in mathematical terms. It should be pointed out in this connection that the so-called measurement of qualities, their presumed quantification, is not in fact the measurement of a quality, but of a concomitant quantity precisely. Qualities as such cannot be measured, cannot in any way be quantified.

Now, the point to be made is simply this: The extension of celestial bodies—like the qualities here below—does *not* admit of mutually external parts; it is not divisible. The kind of extension that breaks into separated parts does not exist in the world above; in that realm, extension is intensive, as Hamberger says: it is an extension, one can say, not in "breadth," but in "depth." As we pass from our world to the celestial, what is lost is thus precisely the quantitative aspect of things. The qualities, here below, do also, no doubt, bear an earthly stamp; but even so, they derive their essence from the celestial realm. The qualities are thus like a glass through which the supernal light filters "obscurely" into our world, to use St. Paul's metaphor. The function of quantity, on the other hand, is not in fact to transmit—to convey essence—but to separate, or better said, to externalize. It is thus by virtue of their quantitative aspect that the things of earth are excluded from the inner world; it is on account of its extension in "breadth" that the camel is unable to pass through the needle's eye. And it is therefore the quantitative aspect of things that will be destroyed at the end of time: *"As a vesture shalt Thou fold them up, and they shall be changed."*[21]

The ingredient which bestows upon the things of this world their distinctive materiality, their earthly cast, is thus none other than *quantity*: no wonder we find it hard to let go of our quantitative conceptions, as indeed we must if we are to grasp the nature of celestial corporeality! It can indeed be said that all things in this world are rooted in quantity. Like a plant emerging out of the earth,

21. Heb. 1:12.

they evolve out of a quantitative substrate, a *materia quantitate signata* that underlies our universe. It is this quantitative substrate, moreover, which becomes objectified through the *modus operandi* of physics, and presents itself to the eye of the scientist as the physical universe, his world of atoms and of galaxies, as we have seen in the earlier chapters of this book. It needs to be clearly understood, however, that this physical universe, which constitutes the intentional object of physics, does not simply coincide with our world, but represents only its quantitative aspect. Physics is blind to the qualities, blind therefore to all essence. It sees only what is indeed outermost, and thus what stands at the furthest remove from the ontologic core of the integral cosmos.

But needless to say, this is not how the physicist generally conceives of his discipline. Trained and conditioned, as he is, to be reductive, i.e., to think of the real in rigorously quantitative terms, he habitually identifies the physical universe with the world at large. And even when he turns mystical, as he occasionally does, that mysticism itself tends to be of a reductive kind: if there be no room in the scientistic ontology even for such a thing as color—as indeed there is not—what to speak, then, of celestial realities!

One sees that the concept of celestial corporeality hinges upon the notion of "intensive extension": nothing less than this will do.[22] But if the notion of "spatial extension" needs to be thus reformulated in reference to the spiritual domain, must not the idea of "temporal duration" be likewise shifted from "extensive" to "intensive" mode? We propose to show that such is indeed the case.

No one will be surprised to learn that what takes the place of time in the spiritual world is *eternity*[23]; yet the question remains: what

---

22. I will mention in passing that it is tantamount to the concept of *intensio interna* employed by Francisco Suarez in the theology of the Eucharist.

23. Strictly speaking, we should say "aeviternity." Yet inasmuch as that distinction does not substantially affect what we have to say, one may as well leave it out of account.

then becomes of duration? We generally incline to think of eternity as an endless duration, which however proves to be contradictory, given that a duration is defined by its beginning and its end. Neither of course can eternity be conceived as a finite duration. Theologians have conceived of it as a *nunc stans* or "now that stands," which seems, like the so-called "moment of time," to have no duration at all. Yet a duration of some kind it must have on pain of not existing at all; for in the absence of duration, or of something that replaces it, eternity would be as fictitious as is the aforesaid "moment of time," which in fact does not exist.

A duration, then, it must have; the point, however, is that this duration cannot be temporal, can no longer be "extensive." What is called for, evidently, is a concept of "intensive duration," the temporal counterpart of Hamberger's "intensive extension." Eternity too must be "without separation," which implies that its duration cannot "fly apart" into separated moments as the temporal does. But here again our sense-based imagination fails us. The problem is that we are attempting to grasp an existence—a *life* in fact—which is no longer mortal, but angelic, if not indeed *divine*.

Let me note, in passing, that this explains why Jacob Boehme speaks of the primordial "seven-part cycle" in temporal terms, as if it were a transformation consisting of seven separate parts or phases. The point is that he does so by force of necessity; as Boehme himself explains in his first work, the *Aurora*:

> If I am to render the Birth of God out of Himself comprehensible to you, I must indeed speak in a devilish manner, as if the eternal Light had been ignited in Darkness, as if God had a beginning; otherwise I cannot teach you, so that you will understand.

Perhaps an observation attributed to Mozart—which has indeed the ring of authenticity—may serve to put us on the right track. I am referring to the composer's amazing claim that, in the first moment of artistic inspiration, an entire musical composition can present itself, unbroken, to his inner sense: it is only at a later stage that the work becomes divided into temporally successive parts, terminating finally in a succession of individual notes. These parts and these notes, therefore, must have been somehow comprehended within the original composition as perceived by the inner sense—but not as successive, not as "mutually external"! What confronts us here is evidently analogous to Hamberger's "intensive extension," but now transposed from the spatial to the temporal realm.

All this relates intimately to the crucial question of "life eternal," a life which can be conceived neither as a temporal succession of states, much less as a *stasis* in which nothing transpires at all. The truth demands the negation of both alternatives; for as Boethius has expressed it so beautifully, what it entails is "the perfect possession of an interminable life *all at once.*" What the partisans of a static present fail to realize is that the future as well as the past are mysteriously comprehended within that "now" of eternity. These two remaining "parts of time" are not simply annihilated, but brought into coincidence; as Nicholas of Cusa points out, in eternity "all temporal succession coincides in one eternal *now.*"

Hard indeed as it is to grasp the point of such statements, one thing at least is clear: only thus—only on the basis of such a "coincidence"—can a "perfect possession of an interminable life *all at once*" be conceived. The fact emerges moreover that life alone *can be* eternal—because eternity *is*, finally, life. The life that we know—a life that breaks up into successive periods and terminates in death—is a reduced life, a life no longer whole; to which one should add that this fatal defect arises precisely from what theology terms Original Sin: all life here below—including the subhuman—has thus been maimed. True life, on the other hand, is eternal by nature: "ever resting in its movement and ever moving in its rest, ever new and yet ever the same" as Franz von Baader magnificently says.

One sees, finally, that celestial corporeality, unlike the corporeality of material substances, does not pertain to objects as such, but belongs rather to spiritual beings, be they angelic, human, or indeed divine. What stands at issue is a corporeality expressive of eternal life, without which *there can in fact be no such life at all.* It is indeed exactly as Hamberger declares it to be: these celestial bodies "do not extend outwards, but inwards, not in breadth but in depth; and that depth stands as much above that breadth *as eternity looms above time.*"

5

The Extrapolated Universe

In the present chapter I propose to reflect upon the discrepancy between our so-called scientific cosmology and the Patristic teaching concerning the creation and early history of the world, a doctrine which respects the literal sense of Genesis. One can hardly conceive of two more divergent and indeed antithetical visions of the universe, which of course explains why the Patristic worldview has come to be regarded almost universally as an antiquated and untenable doctrine, at best a kind of edifying myth. As I propose to show, it turns out that the question is not quite so simple.

We need first of all to ask ourselves whether the two divergent visions—the scientific and the Christian—refer indeed to the same cosmos, the same "world"; and surprisingly, one finds that in fact they do not. The key recognition proves once again to be the ontological distinction between the *physical* and the *corporeal* domains, with which by now we have become familiar. On the one hand there are corporeal objects: the things, namely, which can be perceived. Let me reiterate that these corporeal entities belong indeed to the cosmos, the external world, which is to say that from the outset I reject the Cartesian premises—reject what Whitehead terms "bifurcation"—and adopt thus a realist view of sense perception.[1] Having regained, by virtue of this step, the *terra firma* of the traditional philosophic schools, one sees immediately that physical science does not in fact refer directly to the corporeal world; for as we have come to see, it refers ultimately to fundamental particles and their aggregates, things which are categorically imperceptible and consequently

1. On this question I refer to Chapter 4 of my book, *Science and Myth*, op. cit.

not corporeal. These particles and their aggregates thus constitute a second ontological domain, which I designate the *physical.*

It appears that, in the course of the twentieth century, science has unveiled an imperceptible and hitherto unknown stratum of cosmic reality. Never mind the fact that this remarkable discovery has been almost universally misconstrued on account of a Cartesian bias, which in effect denies the corporeal: what concerns us in the present inquiry is the fact that there are these two disparate domains—the physical and the corporeal—and that henceforth every cosmological debate shall need, *de jure,* to distinguish between these respective "worlds." Is it conceivable, then, that the *corporeal* world does in fact accord with the data of Genesis, that is to say, with the Patristic cosmology? I shall argue that this is indeed the case.

The very possibility of a mathematical physics hinges upon the fact that every corporeal object X is associated with a physical object SX from which X formally derives its *quantitative* attributes.[2] One finds, moreover, that X and SX are spatio-temporally coincident, which is to say that the "here and now" of one applies to the other as well, a principle referred to as spatio-temporal continuity. The corporeal domain, in its spatio-temporal totality, corresponds therefore to a certain subset C of the physicist's space-time, a notion which can be made precise.[3] This subset determines moreover a complementary region, which I shall refer to as the *extrapolated universe.* It is mathematically conceivable, of course, that the subset C coincides with space-time at large, which would mean that the complementary set is empty. It is however likewise conceivable that

2. *The Quantum Enigma,* op. cit. See especially pp. 33–36.
3. The problem is this: If P and Q are points in C and R is a point "between" P and Q, one wants R to belong to C as well. Now, in a "flat" space-time, one can define C to be simply the smallest convex subset containing all corporeal objects. The general case of a "curved" space-time (if such exists) is of course more complicated, but can be dealt with as well.

C constitutes but a speck within the space-time of physical cosmology, which would mean that almost the entire universe affirmed in contemporary astrophysics would be in fact extrapolated.

What then are the true spatio-temporal dimensions of C? How far, starting from the verified "here and now," does the corporeal world extend into the postulated billions of years and light years? Now, the first and most basic point that needs to be made in that regard is that physical science as such is unable *in principle* to answer this question. And the reason for this incapacity is simple: physical science, by its very nature, is restricted in its purview to the physical domain. What it "sees" through its man-made lenses are physical entities and nothing else; and that is moreover the reason why, on a fundamental level, the act of measurement—which necessarily involves a transition from the physical to the corporeal—presents itself to the physicist as an enigma, an anomaly bordering upon paradox. Physics is unable to ascertain the dimensions of C for exactly the same reason why it cannot resolve the enigma of state vector collapse: the fact, namely, that *physics has eyes only for the physical.*

What, then, can one do? If physical science is unable to enlighten us regarding the spatial and temporal extent of the corporeal world, by what means can this knowledge be gained? By what means, in particular, can we ascertain the *age* of that world? It was St. Thomas Aquinas, let us recall, who raised the question whether reason alone could demonstrate that the world had a temporal beginning, and found that in fact it could not, and that it is only by means of Revelation that the issue can be resolved. So too, one may ask, when it comes to the *age* of the corporeal world, could it not be Revelation, once again, that holds the key? One knows that Scripture, taken at its word, does in fact answer the question. It informs us that the world is *young*, which is to say that its age is measured—not in millions or billions—but in *thousands* of years. The exact number, of course, is hardly at issue, and may indeed be in doubt; there are in fact discrepancies in that regard between the Septuagint and Massoretic texts. But the order of magnitude—which is all that presently concerns us—is certainly *not* in doubt. And so, on this basis, one arrives at the conclusion that the region of physical space-time corresponding to the *corporeal* world constitutes indeed a mere speck within the big-bang cosmos, with its billions of years and light years. Except for that "speck," the universe of contemporary cosmology appears on scriptural grounds to be in fact extrapolated. We shall presently reflect upon the ontological status of the extrapolated cosmos; but first it behooves us to take a closer look at the Patristic doctrine. We need to understand in greater depth what Christianity teaches regarding the nature and history of God's creation.

Before all else it should be pointed out that "free" interpretations of biblical texts are bereft of authority: they amount to little more than private opinions. It is needful to interpret Scripture in light of tradition, or as the Orthodox say, "with the mind of the Fathers," which is indeed, in a mystical sense, "the mind of Christ." Revelation, therefore, resides not simply between the covers of a book, but within the living ambience of the Church, which is truly the Mystical Body of Christ.

My second point is this: Tradition, admittedly, condones multiple interpretations of biblical texts; however, it is a universal principle of traditional exegesis that the literal or direct sense must be respected. And nowhere, let us add, is this principle more crucial than when it comes to the early chapters of Genesis, where the temptation to discard the literal sense becomes nowadays acute. To quote what St. Thomas Aquinas states explicitly in *Summa Theologiae* I.68.1 with reference to the interpretation of Genesis:

> In discussing questions of this kind, two rules are to be observed, as St. Augustine teaches (*Gen. ad lit.*, 1.18). The first is to hold to the truth of Scripture (*veritas Scripturae*) without wavering. The second is that since Holy Scripture can be explained in a multiplicity of senses, one should adhere to a particular explanation only in such measure as to be ready to abandon it, if it be proved with certainty to be false. . . .

As Etienne Gilson explains: "St. Thomas is here in full agreement with St. Augustine, and expressly claims to have taken from him this double principle," beginning with the precept "to maintain steadfastly the literal truth of Scripture."[4] It appears that even in its Scholastic mode, Christianity gives preference to the literal or non-metaphoric sense: that is indeed the *veritas Scripturae* which is to be embraced "without wavering." On the other hand, it can also be said that Latin Christianity has at times exhibited a predilection for metaphorical interpretations of biblical texts—not as contradicting the literal sense, but as being "even more true" if one may put it thus. To cite a major example, the interpretation of the Six Days: the Eastern Fathers—from the Cappadocians to St. Chrysostom, St. Symeon the New Theologian, St. Gregory the Sinaite, right up to St. Seraphim of Sarov and St. John of Kronstadt, who reposed in our century—have consistently emphasized the direct interpretation of these "days" as successive periods of short duration, paradigmatic at least of *our* days, in which the stipulated acts of creation took place. Many of the Latin Fathers, on the other hand, beginning with St. Augustine, have opted for a more metaphysical view of creation, according to which the creative act is one and instantaneous, while the Six Days refer to successive phases in the temporal manifestation of created beings. One could say that, viewed *sub specie aeternitatis*, the six creative deeds of the *hexaemeron* coincide in the *omnia simul* of a single indivisible Act. Whether there be six acts or one depends thus on the point of view: whether one assumes a cosmic or a metaphysical perspective. And I would add that when it comes to cosmology, properly so called, it is the former that takes precedence. So too it is worth noting that St. Ambrose, the teacher of St. Augustine, in a treatise entitled *Hexaemeron,* stands solidly with the Eastern tradition not only in the way he views the Six Days, but indeed on all major points relating to the interpretation of Genesis. It behooves us now to consider this Patristic and biblical doctrine of creation and early history. I shall follow the lead of Hieromonk Seraphim Rose, whose pioneering monograph on this subject has proved invaluable.[5]

---

4. *The Christian Philosophy of St. Thomas Aquinas* (Notre Dame, IN: University of Notre Dame Press, 1994), p. 467.

5. *Genesis, Creation, and Early Man* (Platina, CA: St. Herman of Alaska Brotherhood, 2000). I am profoundly indebted to Fr. Seraphim for opening my eyes to

It is crucial to reiterate, first of all, that according to the Fathers the Six Days are indeed *days*. The notion that these "days" stand in reality for vast periods of time—a position currently known as the day-age theory, progressive creationism, or old-earth creationism—is altogether foreign to the Patristic mind. It needs however to be pointed out from the start that despite their adherence to the direct or literal meaning of Scripture, the Fathers are by no means "fundamentalists" in the contemporary sense. So far from being simple-minded or naïve, I would argue that their doctrine is in fact more subtle—more refined, one could say—than any of the positions commonly espoused in our day. One must bear in mind that many of these Fathers were mystics of a high order, who may indeed have had access to realms that antedate the Fall. For it goes without saying that there must indeed be a categorical divide separating the primeval and the contemporary universe, which ordinary mortals are unable to cross. The herbs and trees, for instance, said to have flourished in Paradise, are by no means the same as the plants existing in our day; and yet they are truly plants, the forerunners of the herbs and trees we know. What Genesis teaches is that the Fall of Adam has radically changed the world itself: *our* world, that is to say. The catastrophe of man's alienation from the Source of his being has affected the cosmos at large. An ontological diminution has taken place on a cosmic scale which we, with our likewise diminished faculties, are simply unable to grasp. The Patristic reading of Genesis, literal though it be, is therefore mystical as well: it *must* be, on pain of missing the mark! Only when that "supernatural" conception of the Edenic world is replaced by a profane understanding does the literal interpretation turn fundamentalist; and only then, let us note, does it become vulnerable in principle to attack on scientific grounds.

The history of the world, according to the Patristic teaching,

the true significance of the Patristic cosmology, and for providing ready access to the relevant source material. For the convenience of the reader I will include page references to Fr. Seraphim's treatise (using the abbreviation GCM).

breaks into major segments which need to be clearly discerned and carefully distinguished. The first of these periods, to be sure, is given by the Six Days: the days in which the world itself began. This period is not yet history, properly so called, but rather the beginning of history; and as Fr. Seraphim points out: "What *is* the beginning of all things but a *miracle!*"[6] By means of suitable hypotheses, contemporary science has succeeded to its own satisfaction in concentrating that miracle at a single point, known as the big bang. Yet Genesis informs us that the miracle was spread out, as it were, over Six Days. It is true that "*He who liveth in eternity created all things at once*"[7]; yet even so, viewed from the side of the cosmos that single Act breaks into six successive Acts which together make up the *hexaemeron*, the "work of creation." The Six Days, therefore, in which the world began, are profoundly different from all succeeding days, and together constitute a unique and incomparable period of prehistory.

Following upon the Six Days there ensued what could be termed the Age of Paradise, a period which came to an end with the Fall of Adam. And as we have noted before, the Fathers insist that the Genesis account of that period be understood in a literal and mystical sense at once; for it happens that with the Six Days we have still not arrived at *our* world: the observable universe in which we presently find ourselves. To be precise, the Patristic understanding of Paradise is mystical in two respects: first, because it insists that the nature of Paradise exceeds categorically what the "carnal man"—St. Paul's *psychikos anthropos*—is able to comprehend; and secondly, because it claims that the things of Paradise can in fact be "seen" in certain states of contemplation. St. Gregory the Sinaite, for instance, speaks of this explicitly when he explains "the eight primary visions" accompanying the state of perfect prayer, where things previously hidden "are clearly beheld and known by those who have attained by grace complete purity of mind."[8] And I would add that there *are* in fact no other means by which the things of Paradise can be

6. GCM, p.197.
7. Ecclus. 18:1.
8. Quoted in GCM, p.416.

known: the purified mind constitutes the only "telescope," the one and only "scientific instrument" by which these things can be brought within range of human observation.

We learn from Genesis, and from the Fathers, that in the Age of Paradise the state of the world and the conditions of life on Earth were vastly different from what we know today. For example, "*the wild beasts of the earth,*" and "*all flying creatures in heaven,*" and "*every reptile creeping on the earth which has in itself the breath of life*"[9] were as yet what we would term "herbivorous." And as to the nature of Adam, the Fathers accept without question that "*God created man incorruptible.*"[10] As St. John Chrysostom observes: "*Man lived on earth like an Angel; he was in the body, but had no bodily needs.*"[11] It appears that by contemporary standards the Age of Paradise is still miraculous. Only after the Fall, and in consequence thereof, do the contours of *our* world begin to come into view; for indeed, "*By man came death.*"[12] Think of it: where are those Darwinist "chains of descent" entailing death upon death over millions of years! This is a point of capital importance: whosoever does not accept the dogmatic claim "*By man came death*" has *ipso facto* rejected the Christian worldview. From that point onwards nothing fits anymore: the entire Christian doctrine becomes unraveled. Once that pivotal contention—"*By man came death*"—has been categorized as an absurdity, the Incarnation, the Redemption, and the Resurrection follow suit. On this point Teilhard de Chardin was not mistaken in the least: given that the evolutionist account of origins is true, it is needful that the doctrine of Christianity be radically revised: "moved to a new foundation" as he said. One has only two options: to stand with Scripture and the Fathers of the Church, or cast one's lot with the Jesuit paleontologist and his successors; a middle ground does not exist.[13]

Let us continue. *Our* world—the universe as we know it or nor-

9. Genesis 1:30.

10. Wisdom 2:23.

11. GCM, p. 444.

12. 1 Cor. 15:21.

13. On this subject I refer to my monograph, *Theistic Evolution: The Teilhardian Heresy* (Tacoma, WA: Angelico Press/Sophia Perennis, 2012).

mally conceive of it—came into being *after* the Fall. A cataclysmic transformation of unimaginable proportions has radically altered not only the external world, but man himself, beginning with his cognitive faculties. As St. Macarius the Great informs us, the bodily expulsion from Paradise had its counterpart in the soul: "That Paradise was closed," he writes, "and that a Cherubim was commanded to prevent men from entering it by a flaming sword: of this we believe that in visible fashion it was just as it is written, and at the same time we find that this occurs mystically in every soul."[14] It was at this juncture that the *psychikos anthropos*, the carnal man *"who receiveth not the things of the Spirit of God: for they are foolishness unto him,"* and knows them not *"because they are spiritually discerned"*[15]—that man *as we know him!*—came to be. Not instantly, in fact, but gradually; for as we also learn from Genesis, Adam and Eve remained for some time "close" to Paradise: close enough to see it from afar. It can be said, moreover, that since the original Fall mankind has been engaged in an ongoing fall from Paradise: every betrayal, large or small—everything that theology knows as "sin"—constitutes a link, as it were, in this (decidedly non-Darwinist!) "chain of descent."[16]

There is reason to believe that even the so-called laws of nature as we know them came into force with the Fall; St. Symeon the New Theologian, for example, suggests this quite clearly when he writes: "The words and decrees of God become the laws of nature. Therefore also the decree of God, uttered by Him as a result of the disobedience of the first Adam—that is, the decree to him of death and corruption—became the law of nature, eternal and unalterable."[17] We need however to understand these last two adjectives in a relative sense; for surely St. Symeon understood well enough that these "eternal and unalterable" laws will again be suspended on "the last

14. GCM, p. 405.

15. 1 Cor. 2:14.

16. I have dealt with this question at some length in *Cosmos and Transcendence* (Tacoma, WA: Angelico Press/Sophia Perennis, 2008), chapter 7.

17. *The First-Created Man* (Platina, CA: St. Herman of Alaska Brotherhood, 1994), pp. 82–83.

Day," when "*the powers of the heavens will be shaken*"[18] and there will be "*new heavens and a new earth; and the former shall be remembered no more...*."[19] It appears that what we know as the laws of nature—what the physicist has his eye upon—applies to an interim phase of the cosmos: the period, namely, between the Fall and what theology knows as the general Resurrection.

The consequences of this Christian claim are of course incalculable. Inasmuch as the Earth and the heavenly bodies, together with the primary forms of life, including man, came into existence *before* the Fall, the evolutionist scenario—from the big bang to the formation of galaxies, stars and planets, and thence to increasingly complex molecules, culminating in biological species—this entire scenario, I say, has been cut off at one stroke. To the extent that we have understood "the mind of the Fathers" we realize that the Six Days and the Adamic world before the Fall transcend categorically what "science"—and indeed the *psychikos anthropos* as such—can ascertain. One does not need to examine in detail the backward extrapolations by which scientists have sought to reconstruct the distant past—right up to the stipulated big bang, and of late, even beyond!—because one knows by way of Scripture and tradition that past a certain point, and well before one exits the integral domain of human history, these extrapolations are bound to be fictitious. One therefore knows from the start that assumptions of a non-verifiable nature must have been smuggled into these extrapolations, and that the observable facts and tested laws of nature do not suffice to yield the evolutionist scenario. We shall return to this question presently.

But first it is needful to observe that Genesis speaks of yet another major turning point in the history of the world, by which the Earth as we know it came to be. I am referring of course to the Flood, which the Genesis account describes in striking detail. We are told, for example, that the waters rose to a height of fifteen cubits—about twenty-two and a half feet!—above the highest mountain peak: what are we to make of these biblical claims? Theologians, to be sure, have tended for about a century and a half to shy away from a

18. Matt. 24:29.
19. Isa. 65:16.

literal interpretation of the relevant texts, in fear no doubt of ridicule from the scientific sector. The Fathers, on the other hand, taking Genesis at its word, have perceived the Flood as a catastrophe which has drastically altered the face of the Earth as well as the conditions of terrestrial life. With Noah and his descendants a new phase of history—of human *and* terrestrial history—commences: even the climatic conditions appear to have radically changed. Before the Flood, one is led to conclude, the distribution of water on Earth was significantly different from what it is today. Apparently a great quantity of moisture had been diffused throughout the atmosphere, causing what is nowadays called a greenhouse effect. It was presumably in part these "waters above" that came down at the time of the Flood to cause a global inundation. It has been suggested that prior to the Flood direct sunlight could not penetrate to the surface of the Earth, which might account for the fact that the rainbow—the sign of the primordial covenant—appeared for the first time *after* the Flood. One may speculate that during the Flood, and also presumably in its wake, gigantic upheavals in the tectonic structure of the Earth have taken place, by which the present topography came to exist. We must not forget that the Genesis Flood— assuming of course that it occurred—falls perforce within the province of science in the contemporary sense. Unlike the seemingly miraculous events preceding the Fall, it is something that happened some five thousand years ago. Is this a claim, then, which can be reasonably maintained today?

It is to be noted from the start that this is not a matter of merely academic or marginal interest; for in fact the contention is indispensable to the Patristic *Weltanschauung*, which stands or falls as the integral doctrine it is. Here too it can be said: deny a part, and you have denied the whole. I propose now to consider the status of the Patristic doctrine in light of known scientific facts.

Until the early decades of the nineteenth century it was widely believed that the Earth is some seven or eight thousand years old, and that the major fossiliferous strata were indeed deposited at the time of the Flood. By 1900 the accepted age of the Earth had grown to about 100 million years, and today it is five billion, give or take. And to be sure, it is now the generally accepted view that fossiliferous strata were formed over vast periods of geologic history by a more or less uniform process of sedimentation. The shift began in 1830 when Charles Lyell announced the principles of uniformitarian geology, based upon the assumption of an "old Earth." Nine years later John Pye-Smith took the next step: declaring the Genesis Flood to have been a local inundation in the region of Mesopotamia, he eliminated in effect the only viable alternative to Lyell's theory. The stage was now set for Charles Darwin, whose *magnum opus* appeared in 1859.

How, let us ask first of all, does one substantiate the old-Earth hypothesis? The principal means of estimating geologic age is by radiometric dating of igneous and metamorphic rocks. The idea is simple. Given that a sample contains traces of a radioactive isotope plus some element belonging to its decay series (commonly referred to as parent and daughter elements, respectively), one can calculate the length of time needed to produce the observed ratio of the two—*provided* one knows the initial ratio, and knows also that no contamination or leaching has taken place during the process of decay. And of course there is a third generic assumption: namely, that the rate of radioactive decay some millions or billions of years ago was the same as it is today. It happens, however, that each of these assumptions is open to serious doubt, especially the first, which seems to be drawn out of thin air precisely to permit radiometric dating. One assumes that initially only the parent element was present; but why should this be the case? One knows, moreover, that usually one or more of these generic assumptions are in fact invalid, for the simple reason that different methods of radiometric dating applied to one and the same sample generally yield significantly different results. In addition, it has been found that radiometric dating, applied to lava samples of known age, can overestimate by incredible factors.

To complicate matters further, it happens that the sedimentary rock which makes up the fossiliferous strata cannot be dated radiometrically at all. It may surprise the non-specialist to learn that these strata are normally dated by means of so-called index fossils, and thus on Darwinist premises! As Edmund M. Spieker—a respected geologist and strict uniformitarian, no less—observes with reference to the time-scale associated with that famous "geologic column" displayed in every natural history museum:

> And what essentially is this actual time-scale, on what criteria does it rest? When all is winnowed out, and the grain reclaimed from the chaff, it is certain that the grain in the product is mainly the paleontologic record and highly likely that the physical evidence is the chaff.[20]

It appears, moreover, that the hypothetical column was "frozen in essentially its present form by 1840," at a time when geologists had examined only bits of Europe and the eastern fringe of North America. "The followers of the founding fathers," writes Spieker, "went forth across the earth and in Procrustean fashion made it fit the sections they found, even in places where the actual evidence literally proclaimed denial. So flexible and accommodating are the 'facts' of geology." Even so-called disconformities—old strata piled upon young—do not seem to disturb these self-assured geologists: as in the case of Darwinism, so in geochronology it appears that problems can be invariably fixed by adding yet another *ad hoc* hypothesis.

The most embarrassing fact, perhaps, is that the stipulated uniformitarian process of fossil formation seems not to be operating anywhere. As Richard Milton points out:

> Today there are no known *fossiliferous* rocks forming anywhere in the world. There is no shortage of organic remains, no lack of quiet sedimentary marine environments. Indeed there are the bones and shells of millions of creatures available on land and sea. But nowhere are these becoming slowly buried in sediments and lithified. They are simply being eroded by wind, tide, weather and predators.[21]

20. Quoted in J.C. Whitcomb and H.N. Morris, *The Genesis Flood* (Phillipsburg, NJ: P&R Publishing, 1998), p. 211.

21. *Shattering the Myths of Darwinism* (Rochester, VT: Park Street Press, 1997), p. 78.

What is evidently needed for fossil formation is *rapid* burial; and as Milton goes on to note: "Not even the most dedicated Darwinist could believe that the average rate of sedimentation of the uniformitarian geologic column (0.2 millimeters per year) is capable of providing rapid burial."

But the worst was yet to come: since 1985 the case for uniformitarian geology has deteriorated dramatically. What happened is that a French geologist by the name of Guy Berthault initiated controlled experiments designed to ascertain the actual mechanism of sedimentation, the results of which seem to disprove the assumptions upon which Lyell's theory is based. "These experiments," writes Berthault, "contradict the idea of the slow buildup of one layer followed by another. The time scale is reduced from hundreds of millions of years to one or more cataclysms producing almost instantaneous laminae."[22] Berthault's results, published between 1986 and 1988 by the French Academy of Sciences, have not unreasonably been referred to as the "death knell" of uniformitarian geology.

Perhaps these few indications, sparse though they be, will suffice to show that geochronology is not in fact the "hard" science one generally takes it to be, and that, when all is said and done, the old-Earth tenet remains today what it has been from the start: an unproved and indeed unprovable hypothesis. An unbiased observer cannot but agree with Milton (himself by no means a young-Earth advocate) when the latter concludes: "Because radioactive dating methods are scientifically unreliable, it is at present impossible to say with any confidence how old the Earth is."[23]

Meanwhile creationists—long disdained and ostracized by the scientific establishment—are conducting respectable geologic research of their own, based upon *their* assumptions, which to be sure are inspired, not by dreams of evolution, but by biblical tenets. The hypothesis of a universal Flood as described in Genesis plays of course a central role in these investigations, and appears today to accord far better with the geologic facts than the opposing uniformitarian premise, which can now be safely written off as a discred-

22. See Milton, ibid., p. 78.
23. Ibid., xi.

ited surmise. Today one no longer needs to be a biblical believer to place one's bet on Flood catastrophism.

It is important to note that the Genesis cosmology is undeniably geocentric. We are told in the very first verse that the Earth was created "in the beginning," before the Sun, Moon, and stars were made. So too one learns that the heavenly bodies were created on the fourth Day, *after* the Earth had been rendered habitable and plant life had appeared. And when God did create "the lights in the firmament of heaven," He made them "to divide the day from the night," and "for signs and seasons." We are given to understand that the cosmos is indeed centered upon the Earth. This preeminence, moreover, mirrors evidently the preeminence of man, the theomorphic creature who, as St. Symeon has expressed it, "was placed by the Creator God as an immortal king over the whole of creation."[24]

The centrality of the Earth in biblical cosmology is first of all *iconic*: we are not yet in the domain of measurable quantities and strictly geometric relations. Yet it is also true, of course, that the prevailing cosmography was in fact geocentric till the advent of modern times, after which the centrality of the Earth came to be progressively diminished or obscured, an historical fact which may not be without spiritual significance. As the matter stands today, the Earth has forfeited every trace of its erstwhile preeminence: neither in our solar system, nor in our galaxy, much less in the cosmos at large is its position presently conceived to be distinctive in any way. We must not forget, however, that geocentrism refers in the first place to the *corporeal* as opposed to the physical world; it has to do thus with the position of the Earth within the previously defined subset C, precisely. And inasmuch as physical science is in principle unable to ascertain the dimensions or boundary of C, it is *ipso facto* incapable of resolving the question of geocentricity as well. Contrary to the prevailing opinion, therefore, geocentrism has not been ruled out of court.

24. *The First-Created Man*, op. cit., p.90.

As to the rival concept of heliocentrism, it is debatable whether the Copernican theory as such contradicts the notion of geocentrality. It is a fact that relative to a heliocentric coordinate system the planetary motions assume a particularly simple and elegant form; as Copernicus himself has put it: "Under this orderly arrangement a wonderful symmetry in the universe" comes to light, "and a definite relation of harmony in the motion and magnitude of the orbs, of a kind not possible to obtain in any other way."[25] But does this prove that it is the Earth and not the Sun that moves, or deny the postulate of geocentrality? Copernicus himself makes neither of these claims, and indeed, from a mathematical point of view the question does not even make sense: which body moves, or which is central, depends upon our choice of coordinates. My point is this: whether we perceive geocentrism to be compromised by the Copernican discovery depends ultimately, not on scientific fact, but on our philosophic orientation. The discovery itself does no more than reveal a hitherto unrecognized "relation of harmony in the motion and magnitude of the orbs"—*ad majorem Dei gloriam!* a Christian can say. And as a matter of fact, Pope Clement VII was delighted to learn of the Copernican discovery, and urged the Polish savant to publish his findings.

The conflict over heliocentrism did not erupt until the following century, when Galileo promulgated not simply a mathematical theory, but an entire worldview. As I have pointed out repeatedly, what underlies the scientistic *Weltanschauung* in its entirety is the so-called bifurcation postulate, and this is precisely what Galileo introduced by his distinction between the so-called primary and secondary qualities. One should add that he did so on spurious grounds, for it happens that there is no scientific basis for that postulate, nor *can* there be. One sees in retrospect that the famous dispute—which has been thoroughly misrepresented in our schools and universities!—was not simply about astronomical facts, as one likes nowadays to believe, but that, underneath the surface, a much larger question was at stake, an issue which in truth vitally affects the Church as the guardian of authentic Christian doctrine. And let me say, in passing, that on the whole the position of the ecclesiastic

25. *De Revolutionibus* 1:10.

authorities, and most especially that of Cardinal Bellarmine, was objective, accurate, and fair in its judgment.[26]

As concerns contemporary cosmologies, the crucial point to be noted is that these are based not simply on empirical facts and known physical laws, as one generally imagines, but require in addition a third ingredient of a very different kind: a cosmological model, namely. But whence are these models derived? How, in particular, does one arrive at the model which underlies the most famous cosmology of our day, the so-called big bang cosmology? Here is what Stephen Hawking and George Ellis have to say on this question:

> We are not able to make cosmological models without some admixture of ideology. In the earliest cosmologies, man placed himself in a commanding position at the centre of the universe. Since the time of Copernicus we have been steadily demoted to a medium sized planet going around a medium sized star on the outer edge of a fairly average galaxy, which is itself simply one of a local group of galaxies. Indeed we are now so democratic that we would not claim our position in space is specially distinguished in any way. We shall, following Bondi, call this assumption the *Copernican principle*. A reasonable interpretation of this somewhat vague principle is to understand it as implying that, when viewed on a suitable scale, the universe is approximately spatially homogeneous.[27]

It appears that geocentrism was cast out, not on the strength of scientific facts, but indeed on *ideological* grounds. The problem seems to be that geocentrism smacks of intelligent design. It happens that the founders and protagonists of astrophysical cosmology espouse an ideology that favors chance or randomness in place of an intelligent Creator; fundamentally, big bang cosmology is Darwinism on a cosmic scale.[28]

---

26. For an in-depth discussion of the geocentrism versus heliocentrism issue, I refer to Chapter 8.

27. *The Large Scale Structure of Space-Time* (Cambridge University Press, 1973), p. 134.

28. We shall have occasion to recur to this point in the chapters to follow. See also *Science and Myth*, op. cit., pp. 194–200.

The question arises now whether one can construct a viable relativistic cosmology starting with a *geocentric* model. And a second question is this: In such a geocentric universe, would the Earth be young or old? To be sure, the prospect of a tenable young-Earth cosmology seems rather slim. Astronomers, we are told, have observed galaxies 12 billion light years distant from our planet: according to biblical chronology, this would mean that the photons responsible for these observations embarked upon their cosmic journey some 12 billion years *before* the world began! On the face of it, the scenario seems absurd. One must remember however that time, relativistically speaking, is not the absolute we normally take it to be, but can run at vastly different rates. It is therefore indeed conceivable that 12 billion years of star-time translates into some thousands of years as measured by terrestrial clocks. Given that biblical chronology refers to Earth-time, there would in that case be no contradiction between a *young* creation and *ancient* stars. But the question remains of course whether this conceptual possibility can be realized in an actual relativistic cosmology.

This brings me to the remarkable investigations of Russell Humphreys, a physicist with biblical as opposed to Darwinist persuasions, who did recently propose a relativistic cosmology along these lines.[29] Humphreys postulates a bounded spherical space with the Earth at its center. The model is classical: Ptolemaic almost. But it is also profoundly Christian: the very geometry of such a universe is indicative of design, of purpose, even as the opposing model of an unbounded universe with an "approximately spatially homogeneous" mass distribution is suggestive of randomness, of blind chance. Boundedness itself, moreover, is suggestive of *transcendence*, a notion closely allied to that of intelligent design, and equally unpalatable to the scientific mainstream. Humphreys nevertheless concurs with the big-bang cosmologists regarding cosmic expansion: this tenet, he believes, is based, not on the so-called Copernican principle, but indeed on empirical grounds. His task, then, is to

29. *Starlight and Time* (Green Forest, AR: Master Books, 1994).

construct an expanding spherical cosmos satisfying the equations of general relativity, a problem, it turns out, that relates intimately to the theory of black holes. A black hole, it will be recalled, is a region of space in which the gravitational field is so strong as to cause both matter and radiation inside the region to be permanently trapped: not even light can escape, hence the term "black hole." Such a region is bounded by an intangible surface called the event horizon, at which light rays emanating from inside the black hole are turned back. The event horizon, moreover, proves to be semi-permeable, which is to say that matter and radiation can pass through that surface in one direction: in this instance, from outside in. One may recall that Stephen Hawking has provided an impressive description of an astronaut traveling toward the event horizon of a black hole;[30] what is striking is the extreme gravitational time dilation[31] experienced as the event horizon is approached: the astronaut's clock slows down (relative to *our* clocks) and comes to a stop as the event horizon is crossed.

Now, it happens that the equations of general relativity allow also a second scenario, which can be briefly characterized as a black hole operating in reverse. One has once again a region of space bounded by a semi-permeable event horizon; but this time matter and radiation can only pass *out* of the region. These strange solutions were discovered mathematically in the seventies and dubbed "white holes," but up till now have proved to be of little interest to astronomers and cosmologists. It turns out, however, that a white hole is precisely what one needs to arrive at a cosmological model that is spatially bounded. Humphreys shows in fact that a spatially *bounded* universe *must have* expanded out of a white hole. In the history of a geocentric universe there must consequently be a moment when the event horizon passes through the surface of the

---

30. *A Brief History of Time* (London: Bantam Books, 1988), pp. 87–88.

31. The so-called special theory of relativity predicts a "velocity time dilation": the rate at which clocks in motion "tick" relative to a reference frame conceived as stationary diminishes with their velocity and tends to zero as one approaches the speed of light. The general theory of relativity, which takes account of gravitational fields, predicts in addition a second kind of time dilation, which is to say that gravitational fields likewise slow the rate of clocks.

Earth; what does this mean? In place of an astronaut approaching the event horizon from the outside, imagine an observer on Earth emerging from inside the white hole. Now it is Earth-time that almost stands still, while the clocks on distant stars seem by comparison to be racing at a fantastic rate: fast enough, in fact, to measure out a billion years in a period of arbitrarily short duration as registered by *our* clocks. Remarkably, it appears that *a geocentric cosmology is perforce a young-Earth cosmology as well.*

It is surely of great interest that such a young-Earth cosmology has been shown to exist, and creationists are justifiably hopeful that it will accord with the observable facts better than the big-bang theory, which actually leaves much to be desired in that regard. We must not forget, however, that physical cosmologies, no matter how well they may fit the empirical data, are subject to a generic limitation arising from the ontological discrepancy between the physical and the corporeal domains; and this is what needs now to be clarified.

First of all I contend, once again, that the corporeal has ontological primacy over the physical. What does this mean? Obviously physical entities exist as intentional objects; but so do the entities perceived in a dream or a mirage. I am by no means suggesting that physical entities are no more than figments of our imagination! My point, rather, is that they are *derived* in relation to the corporeal: *primary* existence, I maintain, stems directly from the creative Act of God, and thus from what St. Thomas Aquinas terms *esse*, the act-of-being itself. As we read in the prologue of St. John's Gospel: *"All things were made by Him, and without Him was not anything made that was made."* Now, *"what was made"* on the ontologic level of our world are *corporeal* entities: all the rest—from a mirage to the particles and fields of physics—hinge upon the corporeal realm, from which they derive such being or reality as they have. A reality they do of course have; but it is a *derived* reality, from which, moreover, the corporeal can *not* in turn be derived.

This position is of course the exact opposite of the reductionist view, which in effect assigns ontological primacy to the physical.

The corporeal comes thus to be conceived as an epiphenomenon of the physical, notwithstanding the fact that this presumed epiphenomenon proves to be scientifically incomprehensible: for indeed, if it were not, it would *ipso facto* reduce to the physical. It is moreover worth noting that the claim of physical primacy is itself opposed to the canons of scientific empiricism, and thus to the dominant trend in the philosophy of science, which is *epistemological* in its fundamental principles: an empiricism, namely, which insists upon the primacy of scientific observation, of controlled sense experience one can say. The claim of physical primacy stands thus actually in contradiction to the philosophic foundations upon which physics itself is based. It was Niels Bohr—and not his realist opponents!—who accords with these founding principles when he declares: "There *is no* quantum world; there is only a quantum description."

I would also note, in passing, that one can hardly affirm physical primacy without reifying the physical by endowing it with corporeal attributes. A fundamental particle, thus, comes to be endowed with a shape and position befitting a marble or billiard ball, contrary to what quantum theory permits us to conceive. In a word, in attributing primacy to these putative particles we spuriously corporealize them—a fact which in its own way affirms again the primacy of the corporeal. As Heisenberg has noted, fundamental particles are not, strictly speaking, "things," but constitute "a strange kind of physical entity midway between possibility and reality."[32] To think of these so-called particles as "things," and indeed as the primary things no less—that surely is naïve, to say the least.

Granting the primacy of the corporeal—granting, in other words, that such reality as the so-called fundamental particles may have derives from the corporeal domain—we need to take note of another fundamental fact, which happens to be so simple and obvious as to be easily missed. I contend next that every derived reality is determined by the nature of the primary entities from which it is derived. We may put it this way: secondary realities are determined by a law affecting the primary. A mirage, for example, stems from a

32. *Physics and Philosophy* (New York: Harper & Row, 1962), p. 41.

law of optics, and the holes in cheese from a law that determines the nature of that substance. So too, I say, does the physical derive from a law which defines the nature of the corporeal. That law, to be sure, is not a physical law; it is the principle, rather, which determines both physical things together with the law or laws that apply to them. Now, it goes without saying that God is the author of corporeal nature and its law; we must not forget, however, that one needs to distinguish between corporeal nature *before* and *after* the Fall. And this brings me at last to my main contention: the law, I say, which founds the physical domain came into force precisely at the moment of the Fall. It can in fact be none other than that "law of nature, eternal and unalterable" of which St. Symeon speaks: the law that came into effect when God decreed "death and corruption." We must remember—hard as it may be for us to conceive—that prior to this cosmic catastrophe there *was no* death, there *was no* corruption. It appears that the nature of "flesh," and thus of what we term "matter," has drastically changed. I would point out that we catch a glimpse of this fact in the Gospel accounts of the Resurrection: for example, by the circumstance that the resurrected body of Christ was able to pass through physical barriers such as walls and doors. The entire domain of the authentically miraculous, for that matter, bears witness to the fact that corporeal natures can in principle be freed from the fetters of physical law: that under appropriate conditions these laws are suspended. And it is of course highly significant that the miraculous abounds within the ambience of mystics and saints: of men and women, that is, who have approached or regained the pristine state, the state of Adam and Eve before the Fall.

What, then, does the primacy of the corporeal imply concerning physical cosmology? It forces us to conclude that the physical domain does not antecede the Fall, and will cease to exist the instant "new heavens and a new earth" shall come to be. No physical cosmology, therefore, retains validity as one extrapolates beyond either of these two God-given bounds of human and cosmic history. Extended beyond either divide, physical theory retains a merely formal sense; in other words, it becomes fictitious. To be sure, such a fiction may yet prove to be of use in organizing or predicting certain facts of observation: the very idea of extrapolation implies as much. One falls into error, however, the moment one regards the extrapolated theory as a factual account of a past or future state. *No physical theory, moreover, remains factual as one extrapolates backward to the formation of the first man.*

This does not mean, however, that a physical theory can convey no information whatsoever past the two given thresholds: the point, rather, is that whatever knowledge it may convey beyond these bounds ceases to be scientific. It will be instructive in that regard to reflect somewhat on white hole cosmology. The physical theory needs evidently to be "calibrated" by identifying the moment at which the shrinking event horizon crosses the surface of the Earth. Humphreys, biblical creationist that he is, situates this pivotal event at the fourth Day to coincide with the formation of the Sun, Moon, and stars; but in so doing, he extends the physical theory backwards into a zone to which, in light of our foregoing considerations, it does not apply. I propose instead that the moment in question be taken to coincide with the Fall, which is in fact precisely the instant at which *physical* time—time as we know it—begins. That moment, therefore, has no *physical* prehistory; in the language of Thomistic metaphysics, it marks the passage from aeviternity to time.[33]

Now, it happens that Humphreys' theory bears witness to this metaphysical fact: for at that very moment—given the proposed recalibration—the gravitational time dilation on Earth becomes

33. Aeviternity, properly so called, may be characterized as the temporality proper to the celestial or angelic order. But whereas Paradise is situated "below" the celestial state and may already be subject to time, it yet partakes of aeviternity; as St. Chrysostom observes: "Man lived on Earth like an angel." And I would observe that Edenic temporality appears indeed to be a mean between aeviternity and secular time, which is moreover recovered or attained in the liturgical act, and above all in the traditional rite of Mass, in which, as theology teaches, temporal separation is definitively transcended. It is in this sense that the moment of the Fall "marks the passage from aeviternity to time": at that instant time ceased to be "liturgical," its link with eternity was broken. And it is by way of this scission that "death and corruption" entered the world.

infinite. For an instant as it were, clocks on Earth stand still: Earth-time (looking backwards) comes to a stop. Physical time itself, moreover, and thus the physical domain as such, springs into being at the very moment when the spherical event horizon drops below the surface of the Earth. The resultant image, thus, of the Earth "bursting through" that surface may indeed be seen as an icon of the Expulsion—despite the fact that it refers to a domain which no longer admits of scientific description.

Meanwhile time in the firmament is racing at incredible speeds relative to the newly inaugurated Earth-time: is there also perhaps a symbolic interpretation—a metaphysical explanation if you will—of this discrepancy in the rate of clocks? It seems to me there is. Consider the fact that the entire movement is centrifugal: away from the Source and Center, and thus directed towards a Periphery, what Scripture so descriptively terms *"outer darkness."* Now, if the slowing of clocks to the point of momentary stoppage be indicative of proximity to the Source, to the Edenic state, does it not stand to reason that the speed of clocks as such is indicative of *distance*—of *ontological* distance to be sure—from that primeval Center? In light of Patristic doctrine this supposition is by no means unreasonable. On this basis, however, one can readily understand the racing of clocks as one moves towards the Periphery. It thus appears that Russell Humphreys may have given us a relativistic cosmology which turns out to be not only metaphysically meaningful, but theologically sound.

Yet fascinating as white hole cosmology may be, it too remains inconclusive. To be sure, it is already of enormous interest that a geocentric cosmology is relativistically conceivable at all. Most amazing of all, however, is the fact that such a geocentric cosmology is perforce a young-Earth cosmology as well: one senses that this must mean something, that a major recognition stands here at issue. Yet the question remains whether, *qua* physical theory, that wonderful cosmology is well founded. What counts, of course, is the empirical evidence, the so-called facts of observation; and let us note that a single contrary finding suffices in principle to bring down even the most attractive theory. Clearly, it is Nature that has the last word.

Yet however the case may turn out, the fact that a relativistic cosmology such as Humphreys' can indeed be conceived illustrates how drastically astrophysical models depend on assumptions which, so far as science goes, are drawn out of thin air. We need moreover to understand that by way of these freely chosen models ideological preconceptions do come into play, notwithstanding the fact that according to the official textbook wisdom the intrusion of ideology is rigorously excluded by the very methodology of science itself. And when it comes to astrophysical cosmology, in particular, that ideological input proves to be crucial, decisive in fact: whether we opt for the so-called Copernican principle for example—as every mainstream astrophysicist, from Einstein to Hawking, has done—or postulate a geocentric and bounded universe, as does Humphreys, leads to utterly different results. Let no one imagine, moreover, that "the facts" will eventually determine which of the two theories is correct: we have no assurance that they will. Nor can it be said that, as the matter stands, the "facts" side with the former: as we shall have occasion to see in the next chapter, the matter is by no means as simple as that. And let us not forget that a cosmological model outside the scientific mainstream can hardly be tested at all, given that those who control the technological means to do so are prone to reject such a theory out of hand, that is to say, on *ideological* grounds.

It behooves us now to take note of the fact that the most profound questions contemporary science has begun to address relate precisely to what happened, or will happen, in the "forbidden domains" defined by the two aforesaid cosmic bounds. To transgress these bounds, and thereby extend the sway of the physical to all that exists—such appears to be the objective of present-day science in its most fundamental inquiries. What stands at issue is a radical inversion in our collective outlook, which began with Galileo and the fall of geocentrism, and has been in progress ever since. But whereas this ostensibly scientific evolution has been operative for many centuries, its ultimate objective has only recently sprung into full view: the publication, in the year 2010, of Stephen Hawking's *The Grand Design*, leaves no doubt in that regard.

The overall picture has now become clear. One sees, first of all, that the postulated astrophysics *extrapolates* in precisely the sense we have assigned to that term. It does so, moreover, not on scientific, but on *ideological* grounds, and thus illegitimately even by scientific standards. But most important of all: the resultant hyperphysics spuriously contradicts the sacred wisdom of mankind, trashes it in fact.

6

The Pitfall of
Astrophysical Cosmology

Rumor has it that the so-called big bang is now "a scientifically proven fact." When Arno Penzias and Robert Wilson discovered the since famous microwave background in 1965, the *New York Times* announced the event with the headline: "SIGNALS IMPLY A BIG BANG UNIVERSE." Whether they do or do not is of course another question; but the fact remains that big bang theory has since become the official cosmology. From that time onwards every science major has been taught to believe that the universe was born approximately fifteen billion years ago in some kind of an explosion, and has been flying apart ever since. He has been told that this is the reason why stars and galaxies are observed to recede with a velocity proportional to their distance, as the American astronomer Edwin Hubble is said to have shown. Meanwhile the image of an expanding "soap-bubble universe," dutifully disseminated by the media, has commended itself to the public at large. A radical transformation of our collective *Weltanschauung* has thus been effected.

The reflections to follow break into three disparate parts. I will first review the current scientific status of big bang theory, and thence proceed to comment upon the new cosmology from a theological point of view. I shall contend that despite its seeming affirmation of a *creatio ex nihilo,* big bang cosmology is in fact profoundly hostile to the Christian faith. In the third part we shall consider the claims of contemporary astrophysics as such, without reference to any particular paradigm.

Big bang theoreticians obviously face a formidable task; the theory is after all obliged to account, at least in suitably rough terms, for the physical evolution of the universe, from what Georges Lemaître termed "the primeval atom" to the vastness of its present state. It is hardly surprising, therefore, that one of the most intensive and protracted research endeavors in the history of science has so far succeeded mainly in exacerbating the difficulties. The story begins, if you will, with Lemaître's "primeval atom" version of big bang theory,[1] presented in 1931 at a scientific conference, and soon rejected by the astrophysics community. Lemaître had anchored his theory to the claim that a big bang was needed to account for the existence of cosmic rays, a conjecture which proved to be mistaken. After a period of inactivity, interest in big bang theory resumed at the end of the Second World War, stimulated conceivably by the spectacle of exploding atom bombs. The second version, in any case, was proposed in 1946 by a charismatic physicist named George Gamov.[2] In place of Lemaître's cosmic rays, Gamov anchored his theory to the chemical elements, which he perceived as a tangible vestige of the big bang. I vividly recall a physics colloquium, at which, to my amazement, Gamov described in detail the nuclear constitution of the universe so many microseconds after the big bang. Even so, his theory also failed: when Fred Hoyle and his collaborators published a paper in 1957 which showed that nucleosynthesis in the interior of stars gives rise to heavy nuclei—in proportions comparable to existing values no less—the second brief era of big bang theory came to an end. For a time it appeared that an alternative cosmology, the so-called steady

1. Georges Lemaître, a student of Arthur Eddington, was a Belgian Jesuit and physicist. His speculations regarding "the primeval atom"—a curious blend of physics and philosophy—seem to have been well received in ecclesiastic circles, judging by the fact that he was soon thereafter appointed director of the Pontifical Academy of Science. The idea of the big bang, oddly enough, goes back to the poet Edgar Allan Poe, who was an avid science amateur. To counter the problem of gravitational collapse, he proposed, in 1849, that the universe came to birth in an explosion.

2. Like Stephen Hawking in *A Brief History of Time*, Gamov commended his vision of the universe to an entire generation in a science best-seller entitled *One, Two, Three, Infinity*.

state theory, had taken the lead—until that historic moment, in 1965, captured by the *New York Times* in its banner headline.

What is the connection, then, between the microwave background and the big bang? It is clear on the basis of fundamental physics that the big bang event (if indeed it did occur) must have produced an abundance of radiation, which moreover must still exist in the universe, for the simple reason, fundamentally, that it has nowhere else to go. Since it must, moreover, exist in a state of thermal equilibrium—again, because there is no "outside" with which it could exchange energy—that radiation is necessarily of a kind emitted by a so-called black body, the temperature of which can be deduced from its frequency distribution. And finally, the radiation field produced by the big bang must be spread evenly throughout the universe, the reason being that homogeneous initial conditions produce homogeneous effects. According to Gamov's original calculations, that radiation field should by now have redshifted to correspond to a black-body temperature of $20°K$, which would place the bulk of it in the microwave range. Gamov's estimate was later revised to $30°K$ by P.J.E. Peebles, and there the matter stood till that day, in 1965, when the microwave background was picked up on a giant "horn antenna" at the Bell Laboratories, and deciphered by two young scientists who had never heard of the big bang. Despite the fact that the radiation turned out to have a black-body temperature of $2.7°K$ (off by a factor of ten), the discovery conveyed the impression that an astounding prediction had now been verified, and that indeed "signals imply a big bang universe."

The theory, however, was not yet in the clear. The biggest problem facing the new cosmology was to account for the large-scale structures of the astronomic universe. And here the microwave background proved to be in fact a formidable obstacle: its very smoothness and isotropy seemed to preclude the kind of "clumpy" universe we now observe. Given that matter in the primeval cosmos was as evenly distributed as the microwave background leads us to believe, how could it become concentrated into stars and galaxies? Presumably some initial fluctuations became amplified under the influence of gravitational forces to form the stellar universe; one finds, however, that the gravitational fields needed to accomplish such a consolidation must be vastly stronger than the total amount of matter in the universe allows. To make matters worse, it turns out that relative velocities between nearby stars and galaxies tend to be far too small to achieve the observed separations within the fifteen or so billion years the big bang scenario allows. Meanwhile the problem has been further exacerbated by a dramatic increase in the dimensions of large-scale stellar objects identified by astronomers. First there were single stars, then galaxies, and then clusters of galaxies; and finally, in 1986, Brent Tully, an astronomer at the University of Hawaii, discovered that most galaxies within a radius of a billion light-years are concentrated into slender structures measuring about a billion light-years in length. These are the so-called superclusters which have since been documented by several research teams. In 1990, Margaret Geller and John Huchra of the Harvard Smithsonian observatory discovered a huge band of galaxies, of supercluster magnitude, which came to be known as the Great Wall; and soon thereafter another team discovered a series of similar structures "behind" that so-called Great Wall. Moving outwards (away from the Earth), they discovered a sequence of "great walls," more or less evenly spaced about six hundred million light-years apart. This is not at all what big bang theorists expected, or wanted to find. In fact, it is about the worst-case scenario, a discovery to which the *Washington Post* responded with another banner headline: "BIG BANG GOES BUST" this time.

Meanwhile a formidable research endeavor has been under way to examine the microwave background with maximum precision in the hope of finding anisotropies. One can well understand why cheers rang out when the COBE (Cosmic Background Explorer) satellite disclosed measurable fluctuations: it was the kind of result the beleaguered theoreticians had been eagerly waiting for. Unfortunately however the variations, measuring around one part in a hundred thousand, proved to be far too small: it appears that fluctuations on the order of one percent are needed to account for the formation of stellar objects such as Tully's superclusters or the Great Wall. The radiation background, it appears, turns out to be too smooth and isotropic to permit a transition from the postulated primeval state to the observed astrophysical universe.

Big bang theorists typically respond to problems of this kind by making additional assumptions. To be sure, a scientific theory need not be instantly discarded when it comes into conflict with facts of observation: it is normal practice to search for an appropriate hypothesis to resolve the impasse, a process which not infrequently leads to further discoveries.[3] But this fact hardly exonerates a theory which enjoys little or no empirical support and is kept alive mainly through a proliferation of such *ad hoc* assumptions. One might object to my second contention on the grounds that big bang theory has in fact predicted the microwave background; not only, however, did it predict a false temperature, but it happens that the microwave background can be cogently explained in other ways.[4] Like Lemaître's "prediction" of cosmic rays and Gamov's of chemical elements, the prediction of a microwave background actually does little to shore up the hypothesis of the big bang. Meanwhile the addition of ever new—and ever more fantastic!—hypotheses to square the original conjecture with the facts is not a good sign; there is point to Brent Tully's candid observation: "It's disturbing to see that there is a new theory every time there is a new observation."

What does one do, for example, when it turns out that the amount of matter in the universe is about a hundred times too small to permit the formation of galaxies? Literally with a stroke of the pen one supplies the missing mass—a hundred times the estimated mass in the universe!—by postulating something called "dark matter": a kind which does not interact with electromagnetic fields and has never yet been observed. A profusion of dark matter candidates has been proposed in recent decades by helpful particle physicists: there are axions, higgsinos, photinos, gravitinos, gluinos, preons, pyrgons, maximons, newtorites, quark nuggets, and nucle-

---

3. Discrepancies, for instance, between observed planetary orbits and trajectories predicted on the basis of Newtonian physics led astronomers to conjecture that these deviations may be caused by an unidentified object. This hypothesis was verified in 1930 with the discovery of the planet Pluto.

4. It can be explained, for example, by the quasi-steady-state theory of Burbidge, Hoyle, and Narlikar (see *Physics Today*, vol. 52, no. 4, April 1999), or by the plasma-physics approach of Hannes Alfvén (*Cosmic Plasma*, Holland: D. Reidel, 1981), both of which appear to be viable alternatives to big bang cosmology.

arites, to mention a few; the problem is that all these wonderful particles exist so far only on paper. But let us suppose that there do actually exist, say, higgsinos or quark nuggets: would this suffice to extricate big bang theory from its quandary? Certainly not; a number of other major problems would remain. What is more, each new hypothesis tends to introduce problems of its own, which will presumably necessitate the introduction of additional hypotheses. There is no reason on earth to believe that this procedure will eventually converge; and if it should, one wonders whether one has then *found* a truth, or *constructed* it, as Sir Arthur Eddington, in company with some postmodernist philosophers of science, maintains.[5]

We have so far considered only one major difficulty: the problem of accounting for the formation of large-scale stellar objects. To round out this brief review, I will touch upon one more predicament. Let us note that when it comes to distant stellar objects, from stars to galaxies and clusters of galaxies, all we have to go on is light emitted by the objects in question and received by telescopes, be they terrestrial or mounted on satellites. By light, of course, we understand electromagnetic radiation of whatever frequency, from radio waves down through the visible range to X-rays and gamma rays. In a word, what we receive are light-particles or photons, each of which defines a position on a photographic plate and carries a frequency: that is all. These are, ultimately, the hard empirical facts; the rest is theory, a matter of interpretation. The radiation received does however carry a wealth of information, some of which proves to be quite unequivocal in its implications. It is known from laboratory experiments that the distribution of frequencies emitted by a chemical element, for example, is a characteristic of that element. The emission spectrum constitutes thus a signature permitting us to detect the presence of hydrogen, helium, and other elements in stars and galaxies. It happens, however, that spectra received from outer space are generally shifted to the left on the scale of frequencies, a phenomenon known as the redshift; what causes this shift? One has long surmised that stellar redshifts constitute a Doppler effect, which is to say that the reduction of received frequencies is caused by

---

5. On that question I refer back to Chapter 2.

a recessional velocity of the source, just as the pitch of a train whistle is lowered when the train is speeding away from us. On this basis stellar redshifts have been interpreted as evidence for an expanding universe: big bang cosmology is founded upon that hypothesis.

Now, it happens that observational results have accumulated for half a century which seem to contradict that assumption. The first bit of bad news for big bang theorists came in 1963 with the discovery of extragalactic radio sources now known as quasars, whose emission spectrum proved to be heavily redshifted, corresponding to recessional velocities approaching the speed of light. It was soon found, however, that these quasars were typically associated with galaxies whose redshifts are normal, that is to say, relatively small. Stellar objects, thus, which according to big bang geometry were supposed to be billions of light years apart, appeared to be close neighbors, which is to say that the Doppler interpretation of quasar redshifts has become suspect. Meanwhile intrinsic (non-Doppler) redshifts have been surmised in other stellar objects, down to the level of stars; as one authority in this field informs us: "Most extragalactic objects have intrinsic redshifts."[6] But this would mean that the hypothesis upon which big bang cosmology is based—the assumption that stellar redshift equates to recessional velocity—has been disqualified.

Certain developments on the theoretical side, moreover, have augmented these misgivings. In 1977 Jayant Narlikar, an astrophysicist, succeeded in generalizing the equations of relativity so as to allow the masses of fundamental particles to increase with time. The theory, it turns out, predicts intrinsic redshifts caused by this variation of particle mass. The idea is simple: the smaller the mass of an electron, the smaller will be the energy it loses in a so-called quantum jump, which however is just the amount of energy given off in the emitted photon. Since the frequency of a photon is pro-

---

6. Halton Arp, *Seeing Red: Redshifts, Cosmology and Academic Science* (Montreal: Apeiron, 1998), p. 95. One of the foremost experts on quasars, Halton Arp apparently became *persona non grata* among his peers in the U.S. when he began openly to question the Doppler interpretation of stellar redshifts. He is now at the Max Plank Institute for Astrophysics at Munich. His book constitutes an invaluable resource in a field in which it is becoming difficult to separate fact from fiction.

portional to its energy, one obtains thus an intrinsic redshift. In place of the redshift-velocity relation underlying big bang cosmology, the Narlikar theory gives us an inverse redshift-age relation, which enables one to interpret existing data in a new key. The Hubble relation, according to which redshift is proportional to distance, can now be understood from the fact that distant stellar objects are observed at an earlier time due to the finite speed of light, and will consequently tend to have smaller particle masses and correspondingly larger redshifts in proportion to their distance. The heavily redshifted quasars, on the other hand, which do not satisfy the Hubble relation, are now perceived as being constituted by recently created particles ejected from an active or so-called Seyfert galaxy. In this way the new theory does justice to all the relevant facts: to those which support the Hubble relation, as well as those which do not. But now the universe is *not* expanding, nor did it originate in a singularity: in a word, there *is* no big bang.[7]

It is not my objective to tout the Arp-Narlikar approach, which presumably has problems of its own. My purpose, rather, is first of all to bring home the fact that in resting upon the Doppler interpretation of stellar redshifts, big bang cosmology stands on demonstrably shaky ground; and secondly, that there appear to be viable astrophysical alternatives which cannot rightfully be rejected out of hand. Yet the obvious fact remains that big bang cosmology stands nonetheless as the official doctrine. Many astrophysicists are no doubt disturbed by the lack of evidential clarity, to put the case mildly, but few at present seem prepared to challenge the *status quo*. Admittedly, a handful of leading scientists have openly proclaimed the demise of big bang cosmology,[8] and even the conservative science journal *Nature* has run a lead editorial under the caption "Down with the Big Bang." But such occasional expressions of dissent have so far had little effect upon the astrophysics establishment at large; too many careers, it seems, hang in the balance.

7. Ibid., pp. 225–233.
8. Foremost among this handful is the late Fred Hoyle, one of the pioneers.

We propose now to look at the big bang scenario from a theological perspective. Leaving aside the question as to whether this cosmology is factually correct, we shall treat it as a kind of myth or icon, a symbol to be read. What, then, does the big bang signify? What above all strikes one is the idea of a temporal origin: the notion that the universe "did not always exist." This is not to say that "long ago" the world did not exist, for time as we know it refers to cosmic events and cannot therefore antedate the universe itself: "Beyond all doubt," says St. Augustine, "the world was not made *in* time, but *with* time." What big bang theory affirms, rather, is that the universe has a finite age; the question, now, is whether this implies an act of creation *ex nihilo*. I would argue that, from a strictly logical point of view, it does not. But this is actually beside the point: we are now "reading the icon," a task which is not simply a matter of logical analysis. In its iconic import, I say, the big bang picture does overwhelmingly suggest what Christianity has always taught: namely, that the universe was brought into being some finite time ago through a creative act. As Pope Pius XII declared in 1951, in an address to the Pontifical Academy of Science:

> In fact, it seems that present-day science, with one sweeping step back across millions of centuries, has succeeded in bearing witness to that primordial *Fiat lux* uttered at the moment when, along with matter, there burst forth from nothing a sea of light and radiation.... Hence, creation took place in time; therefore, there exists a creator; therefore, God exists![9]

It would seem from this animated papal expression of assent that the impact of big bang cosmology upon Christianity is bound to be salutary; but such proves not to be the case. I contend that the new cosmology has in fact exerted a baneful influence upon Christian thought, and has contributed significantly to the deviations and

---

9. As we shall presently see, it is likewise significant that Pope John Paul II, in another address to the Pontifical Academy, delivered in 1988, warned against "making uncritical and overhasty use for apologetic purposes of such recent theories as that of the big bang."

vagaries afflicting contemporary theology; how can this be? The answer is simple: icons can be dangerous, lethal actually, due to the fact that the icon itself can be mistaken for the truth, "the finger for the moon" as the Chinese say. And this is what has actually happened in the case of the big bang: we are dealing, after all, with a scientific paradigm declared by the leading authorities to be factually true. Now, the problem is that in its *factual* as distinguished from its *symbolic* significance, the big bang scenario is flatly opposed to the traditional Christian cosmogony based upon Genesis. Take for instance the biblical fact that the Earth and its flora were created *before* the Sun, Moon and stars: surely this rules out all contemporary theories of stellar evolution, even as it rules out all Darwinist claims. Theologians, as we know, have for the most part responded to this challenge by "demythologizing" the first three chapters of Genesis; but in so doing, I contend once again, they have taken a wrong turn. Placing their trust in a man-made theory, which moreover stands demonstrably on shaky ground, they have contradicted the inspired teaching of the Fathers and the Church. Let it be said once again that the first three chapters of Genesis, taken in their literal historical sense, cannot be denied without grave injury to the Christian faith. The point has already been made implicitly in the preceding chapter: in bringing to light the content of biblical cosmogony, we have at the same time demonstrated its central importance to Christian doctrine. Whatever contemporary theologians may say in their pursuit of "scientific correctness," the fact remains that the teachings of Christianity presuppose the biblical cosmogony, even as the Redemption presupposes the Fall. It is utterly chimerical, thus, to imagine that the doctrine of Christ actually makes sense in a big bang universe; and one might add that the biblical cosmogony has in fact been mandated by the Pontifical Biblical Commission in 1909. In a definitive response to eight questions relating to "The Historical Character of the Earlier Chapters of Genesis" the Commission explicitly denies the validity of "exegetical systems" which exclude the literal historical sense of the first three chapters.[10]

Getting back to big bang cosmology, I would like to point out that

---

10. Denzinger, *The Sources of Catholic Dogma*, 2121–2128. It is to be noted that

this doctrine is evidently all the more compelling to a Christian public on account of its obvious symbolic signification: what could be
more wonderful, after all, than a scientific cosmology bearing witness to the primordial *Fiat lux!* In conjunction with certain other
scientific developments, the new cosmology has thus fostered a
major movement of reconciliation between the scientific and the
religious communities. Book titles such as "God and the New Physics" (by physicist Paul Davies) or "God and the Astronomers" (by
the astronomer Robert Jastrow) have come to abound, and it is
hardly possible, these days, to keep up with the profusion of semi-
nars and symposia on "science and religion" being held all over the
world. And everywhere one encounters the same message of "peace
and harmony" from both of the former contestants. There is however a price to be paid on the part of religion: wherever a conflict
does arise—as between Genesis and the big bang—it is always
Christianity which is obliged, by the presiding experts, to conform
its teachings to the latest scientific theory. It appears that a certain
fusion of science and religion is now in progress on a world-wide
scale, which threatens to transform Christianity into some kind of
"theistic evolutionism" more or less akin to the quasi-theology of
Teilhard de Chardin.[11]

In a word, the new cosmology is not quite as innocuous as one
might think. So far from being compatible with the truth of Christianity, it proves to be one of the most seductive and potentially
lethal doctrines ever to threaten the integrity of the Christian faith:
a dogma amply capable, it seems, of "deceiving even the elect." The
devil, they say, gives us nine truths, only to catch us in the end with
a lie: could big bang cosmology be a case in point? Could *this* be the
underlying reason why an atheistic science has now promulgated—
to everyone's amazement!—a doctrine which, on the face of it, glo-
rifies God as the creator of the universe? It has at times been suggested that there is indeed a connection between the scientific

---

Pope St. Pius X, in his *motu proprio* "Praestantia Scripturae," declared the decision
of the Biblical Commission binding.

11. I have dealt with this question at length in *Theistic Evolution: The Teilhardian Heresy,* op. cit.

enterprise and the demonic realm; this has been seriously affirmed, for example, by the late Orthodox Hieromonk Seraphim Rose, and again by the Catholic historian Solange Hertz. It is not easy, of course, to document such a connection; but the surmise of demonic influence is neither irrational nor indeed improbable. When it comes to a major onslaught against the Catholic faith, it behooves us to recall the sobering admonition of St. Paul, which may well bear also upon the point here at issue: "*Put on the armour of God, that ye may be able to stand against the wiles of the devil. We wrestle not against flesh and blood, but against principalities, against powers, against the rulers of the darkness of the world, against spiritual wickedness in high places.*"[12] The demonic connection then, of which we speak, may prove in the end to be more than a pious fantasy.

If physics be indeed "the science of measurement" as Lord Kelvin declares, it has to do first of all with operational truth: ontological interpretation is secondary and in a way optional. It is a private matter, one might say, which has to do with the philosophic orientation of the particular scientist. In the astrophysical domain, on the other hand, the case is different, for the simple reason that there is little if any operational truth in that realm. There can be no controlled experiments involving distant stars and galaxies! Moreover, the astrophysicist makes few predictions; and when perchance he does, he typically misses the mark by an order of magnitude or two. Unlike "normal" physics, the object of astrophysics is simply to construct a model of the physical universe at large which agrees with the observable facts, i.e., the signals reaching our instruments of detection from the distant vistas of space. It is somewhat like seeking a mathematical formula to fit a pre-ordained set of data points. But normal physics entails incomparably more than that: it carries an operational truth, as I have said, which in fact forms the basis of our technology. Such is not the case with so-called astrophysics, which in fact differs radically from the textbook definitions of physical science.

12. Eph. 6:11–12.

Now, my point is this: a natural science which is not *operational* in its rationale can only be *ontological* in its claims, which is to say that in the case of astrophysics ontological interpretation is no longer optional but primary. In this regard physics "in the large" differs sharply from physics "in the small": from quantum theory, that is, where operational concerns are paramount. When Niels Bohr declared "there is no quantum world," this contention was neither inconsistent with the principles of quantum theory nor abhorrent to the physics community at large; but imagine what the reaction would be if a scientist were to declare "there *is no* astrophysical universe"!

The ontology of astrophysics is of course physical, which is to say that one conceives of stellar objects as "made of" fundamental particles. But why should this ontology be correct in the distant reaches of space-time when it fails (as I have argued repeatedly) in the terrestrial domain? If corporeal objects pertaining to the terrestrial mesocosm prove to be *more* than atomic aggregates, why should stars and galaxies be "nothing but" atomic? What is more, if quantum particles here below do not have an independent existence—if they belong to what John Wheeler terms "the participatory universe"—why should it be otherwise in outer space? My initial contention is that a strictly physical ontology is as fallacious in the stellar domain as it is in the terrestrial sphere of perceptible objects.

I have elsewhere characterized the knowledge of modern physics as "basic but inessential": *basic*, because it refers to the material side of cosmic reality, and *inessential*, because it is incapable of comprehending substantial form whence essence derives. Physics as such cannot know the quiddity or "whatness" of a thing; the very essence of corporeal entities eludes its grasp.[13] Now, if to know a thing is to know its substantial form—to know, in other words, "what" the thing is—then it follows that the knowledge of physics is not a true knowledge. On the other hand, so long as physics remains operational—so long as its theories can indeed be tested by experiment or verified in their technological applications—it evidently does

---

13. I might mention that Jacques Maritain said much the same when he characterized modern physics as "perinoetic."

embody knowledge of a kind: a pragmatic or operational knowledge, namely. The problem, however, is this: Man was made to know, not pragmatically, but in truth. Hence the well-nigh irresistible tendency to *reify* the intentional objects of physics by attributing to them a corporeal nature. Basically one treats the object in question as if it *could* be seen, could actually be touched. A fundamental particle becomes thus a tiny spherical ball, or a kind of wave perhaps, which the mathematically trained can picture. Such kinds of visualization play in fact a necessary and indeed legitimate role in the comprehension of mathematical ideas: the human mind simply cannot function without some sensory support. It is in the domain of physics, however, as opposed to pure mathematics, that this secret art goes astray; for whereas the mathematician understands full well the ontological difference, say, between a function and its graphical representation, the analogous distinction in physics tends to be unrecognized. The reason for this chronic confusion derives no doubt from the fact that the intentional objects of physics are, admittedly, *more* than a mere "thing of the mind," a mere *ens rationis*, even though they are *less* than a corporeal entity, less than a perceptible thing; and clearly this fact imposes demands upon the ontological discernment of the physicist which are not easily met, to say the least. It is hardly surprising, therefore, the phantasmata of sensory representation are routinely projected upon the physical universe, which then becomes quite literally a *fantasy world*.

The *real* world can only be known by way of substantial form, and thus by way of essence; but how can such knowledge be achieved? Strange as it may seem to the modern mind, we can and do know the substantial form of familiar corporeal entities: we know it through cognitive sense perception, to be exact. No use trying to explain that perception in terms of a natural process of whatever kind;[14] as Whitehead has wisely said: "Knowledge is ultimate." Here below, "to know" and "to be" are both ultimate, which is to say that neither reduces to the other. And so we find that cognitive sense perception constitutes indeed a mysterious act, which in a way tran-

14. On this question I refer to my chapter on "The Enigma of Visual Perception" in *Science and Myth*, op. cit.

scends the confines of the natural world. As wise men have long ago pointed out: the eye by which we see is not itself seen.

A major question now presents itself: what about stellar objects? Do we also know *their* substantial forms, their very essence? Can we in fact perceive objects of that kind: do our powers of cognitive perception reach that far? When we see a dot of light in the night sky, are we actually *perceiving* a star or a galaxy? It does not seem that we are. What we perceive is a dot of light, which we may think of in generic terms as a "star." But a star in that sense is precisely something far away and high above, something which categorically exceeds our reach. Actual cognitive perception—the kind that takes place in the terrestrial realm and transcends bifurcation—can hardly be supposed when it comes to the stellar domain. Whatever may be the essence of a star or galaxy, that essence is presumably beyond the reach of human perception; the stars, it turns out, are "above us" not only in a spatial, but also in an ontologic sense. I will point out in passing that *this fact itself justifies and indeed entails the geocentrist claim.*[15]

It appears that the ontological distinction between terrestrial and stellar entities is indigenous to the ancient cosmologies at large. Even St. Thomas Aquinas speaks of stellar substance as "incorruptible," and thus places stellar objects above the category of corporeal entities pertaining to the terrestrial domain. Nor is it mere poetry when St. Paul distinguishes between the two realms in 1 Corinthians 15: "*There are also celestial bodies, and bodies terrestrial: but the glory of the celestial is one, and the glory of the terrestrial is another. There is one glory of the Sun, and another glory of the Moon, and another glory of the stars: for one star differeth from another in glory.*" We need to pay close heed to these words: for given that the intent of this discourse is to distinguish between corruptible and incorruptible bodies, it is clear that St. Paul is speaking in ontological terms. There is an implicit proportionality here: celestial substances are to the terrestrial as what theology terms the resurrected body is to the natural. And to be sure, the latter are ontologically distinct; for whereas the second is "*sown in corruption,*" the first is "*raised in incorruption.*"

---

15. The question of geocentrism will be considered in depth in Chapters 7 and 8.

What could be more indicative of ontological distinction than that? It would consequently be incongruous to suppose that there is no corresponding ontological hiatus between the stellar and the terrestrial realms: that supposition would be irreconcilably opposed not only to the Platonist and Christian ontologies, but indeed to the *sophia perennis* in all its forms.

Among the things in the natural world, the night sky, above all, speaks to us of high and sacred mysteries. Surrounding our earthly realm like an encompassing sphere, it awakens in us a sense of transcendence, an intimation of higher worlds. According to ancient belief, the starlight we see has in fact its source in these supernal worlds. The stars serve thus as a conduit, an aperture, so to speak, in the vault of heaven through which the transcendent light breaks through to illumine the darkness of this nether realm. That celestial light, moreover, illumines not only the external world, but first of all the heart, the intellect of man. I should point out that there is a biblical basis for these ancient beliefs. To begin with, it is profoundly significant that Genesis refers to stars as "lights in the firmament of heaven," suggesting that the quiddity or essence of a star is indeed none other than "light." That stellar light, however, proves not to be primary, even in the order of creation, for we are told that the primary light was created on the first Day. Or rather, it is by virtue of that first-created light that the first Day itself is defined: one must remember that the original creative utterance of God was indeed the *Fiat lux*. The primary light, moreover, constitutes not only the first, but in fact the highest, the most godlike element in creation; it has been rightly called the most direct manifestation of God. And so it is also the highest symbol of God, a symbol hallowed by St. John when he declared that "*God is light, and in Him is no darkness at all.*"[16] The first-created light, however, is not manifest in our world: as the source of all visibility, it exists unseen. Plato implicitly compares that unseen light to the light of the Sun when he refers to the latter as "the author not only of visibility in all visible things, but of generation and nourishment and growth."[17] One

---

16. 1 John 1:5.
17. *Republic* vi.

is strongly reminded of Psalm 35: "*For with thee is the fountain of life: in thy light shall we see light*"! An entire metaphysics of light is concealed in either passage, a metaphysics the Neo-Platonists were eager to unfold. And let us not forget that this doctrine, which is as biblical as it is Platonist, was incorporated into Christianity, notably through the teachings of St. Augustine and the Pseudo-Areopagite.

Getting back to the stellar realm, one sees thus that it constitutes a world of secondary yet supra-physical light, within which the primary light is mysteriously enshrined.[18] One might add that there is a profound connection between the stellar and the angelic realms, a matter which however would take us too far afield. Suffice it to say that a star is incomparably more than a mere aggregate of quantum particles, that it has both an essence and a function which vastly transcend the astrophysical domain. It is equally important, however, to recall that the stars were given to mankind "for signs." Admittedly this biblical affirmation may have an esoteric sense, by which I mean that there may have been a time when men were able to read "what is written in the stars." But it has also, most assuredly, a significance that applies to us all: for as I have noted before, the night sky awakens in us a sense of transcendence, a presentiment of celestial spheres. Even Immanuel Kant, worlds removed as he was from the sapiential traditions, still sensed the grandeur of this cosmic icon; two things, said he, fill the mind with wonder: "the star-spangled sky above and the moral sense within me." How strange that even this prosaic rationalist, whose philosophy is irreconcilably opposed to the *sophia perennis*, could still sense a connection between "the star-spangled sky above" and the "moral law" deep in the heart of man.

---

18. As the quantum-physicist Arthur Zajonc has beautifully said of light as such: "I cannot describe it, my imagination can only just touch its hem, but I know that at its core there seems to live an original 'first light' within which wisdom dwells. A wisdom warmed by love and activated by life" (*Catching the Light*, Oxford University Press, 1995, p.325). I find it truly remarkable that a contemporary physicist should bear witness to the perennial doctrine concerning light, and in such profound and eloquent terms no less! Like the iconic bright spot within the dark field of the yin-yang, it seems that Zajonc compensates for the almost universal nescience of his peers.

It is hardly necessary to point out that this connection has disappeared—has been implicitly denied—in the astrophysical doctrine: that the cosmic icon, set up by the Hand of God, has been replaced in effect by a man-made picture. Yes, what is communicated to the nonspecialist *is* a picture, a kind of image and nothing more; for whatever truth the discipline may enshrine resides perforce in the operational interpretation of a mathematical structure, which is something else entirely. What astrophysics has to offer the public at large is thus quite literally a fantasy world, a kind of science fiction cosmos; and it is this that has come in effect to replace the cosmic icon in educated minds.

As I have explained at length in my monograph on the ontological interpretation of physics,[19] the root error of the contemporary *Weltanschauung* resides in the reification of the physical: the identification, namely, of a corporeal object X with the associated physical object SX. In the astrophysical realm, however, this generic error is compounded by the fact that stellar objects are not, strictly speaking, corporeal, that is to say, perceptible. Whereas in the terrestrial domain the reification of the physical entails thus a single error, in the astrophysical domain it entails two: for here one corporealizes not only the physical, but stellar substances as well. One thus destroys the dimension of transcendence, the verticality of the astronomical "above." The celestial is reduced to the terrestrial, which is then further reduced to the physical; the cosmos is thus homogenized: "democratized" one could almost say.

In thus distinguishing categorically between stellar and corporeal substances, I am adhering to the conception which associates corporeality with cognitive sense perception. It is on this basis, let us recall, that I differentiate between the physical and the corporeal domains. However, one can also speak of corporeality in a wider sense, corresponding to the Vedantic notion of gross (*sthūla*) manifestation. Corporeality in that sense is characterized by the conditions of space and time, and thus includes the stellar world in its spatio-temporal extension. The kind of corporeality to which

---

19. *The Quantum Enigma*, op. cit.

cognitive sense perception gives access can now be qualified as *terrestrial*, the point being that there are modes of corporeality which differ from the terrestrial not just in terms of quantitative or measurable parameters, but in terms of essence.

The thesis of ontological heterogeneity might be contested on the grounds that the universe seems indeed to be very much of one piece: does one not detect the very same spectra of hydrogen, of helium and other elements in the laboratory as well as in the light of stars and galaxies millions of light-years away? Yes, indeed one does; but we must realize that this kind of homogeneity pertains precisely to the *physical* realm, which stands below the level of being, of actual substance. What the physicist perceives, as it were, at the end of his analysis, are aggregates of quantum particles and nothing more; all ontological distinctions are thus obliterated. But what *are* these so-called particles to which everything has been reduced? As Heisenberg has put it, they are indeed "a strange kind of physical entity just in the middle between possibility and reality." The problem with the homogeneous universe of the physicist, it turns out, is that it does not actually exist. At the risk of digression I would point out that there is a lesson to be learned from this, one that applies even in the political and sociological domains: *obliterate ontological distinctions—obliterate hierarchy—and nothing whatsoever remains*; in a word, ontological homogeneity is tantamount to non-existence. But let us get back to the physical universe. At the end of the physicist's analysis, what remains is not one substance, but no substance at all. As Eddington has pointed out, the very idea of substance has no more place in physics—so long, of course, as that discipline is conceived rigorously, that is to say, in its mathematical structures and operational definitions. When the physicist fails to perceive a categorical difference between the substance of a star and the substance of a terrestrial entity, thus, it is because, strictly speaking, he does not perceive any substance at all. Do not, therefore, ask an astrophysicist "*What* is a star?": in his capacity as astrophysicist, he has not the ghost of an idea.

One might add that what I have said with reference to stellar substances applies in principle to planetary bodies as well, beginning with the Moon. To be sure, men have walked on its surface and have brought rock samples back to Earth to be analyzed; and yet I claim, in light of tradition, that lunar substance differs essentially from terrestrial. Two issues are involved: first, the matter of physical observation, and secondly, the new factor of close-up cognitive sense perception. As concerns the first, the preceding observations have made it plain that nothing new, nothing "non-terrestrial," can emerge from an investigation of that nature: what we as physicists find, once again, are aggregates of quantum particles and nothing more. The fact that actual rock samples are now available for chemical analysis does not change the picture: the preceding considerations apply unaltered to this scenario as well. The matter of cognitive sense perception, on the other hand, is not quite so simple, and demands considerations of a very different kind. One needs to recall, in particular, what I have said in the Introduction to this book regarding experiential knowledge of cosmic realities: according to traditional doctrine, a stratum of cosmic reality can be "entered" only by actualizing the corresponding state in ourselves. So long, therefore, as we remain confined to a state corresponding to the terrestrial domain, terrestrial reality is all that we can perceive. To the extent that we are able to perceive lunar substances at all, we are bound therefore to perceive them as terrestrial, which is to say that we do *not* in fact perceive them. Take an animal—or for that matter, a man bereft of culture—into an art museum, and what do they see? What they see inside the museum is basically the same as what they see everywhere else: what exceeds that level is not perceived. Such considerations, of course, do not prove the traditional claims regarding supra-terrestrial substances; they do suffice, however, to deflate the argument of those who maintain that these claims have now been disproved.

It is no doubt true that of all planetary bodies known to us, the Earth alone offers physical conditions capable of sustaining human life. To be sure, from a scientific point of view the physical environments associated with planetary bodies can presumably be explained in terms familiar to us all; and yet, in light of sapiential tradition the prime determinant proves to be essence, the quiddity of these planetary bodies. The physical conditions, it turns out, are neither primary nor accidental, but are linked to the respective essences. We understand this fact well enough when it comes to a living organism, the contours and physical characteristics of which are naturally expressive of its species; but even here we believe in the primacy of physical explanation, which is the reason why we are committed to an evolutionist biology: in a de-essentialized cosmos physical parameters are all that is left. When essences come into play, on the other hand, it becomes possible to understand the recognizable facts in an altogether different way, through what could be termed a "top-down" approach to cosmic reality. On this basis it becomes clear, in particular, that mankind finds itself on Earth, not on account of some physical contingency, but by virtue of a profound kinship. As I have noted before, it is on account of this inner kinship that we are able to "enter" the terrestrial stratum of cosmic reality by way of cognitive sense perception: the miracle of perception, I say, hinges upon a conformity of essence. On the other hand, as the matter stands there is no such kinship between mankind and the Moon or Mars; and I would add that it is not an accident, therefore, that even the physical ambience of these planetary bodies proves hostile and indeed lethal to man. Strictly speaking, contemporary cosmology is misnamed, because in truth it knows nothing of a *cosmos*, that is to say, an *ordered* world.

Getting back to what might more properly be termed global cosmography, I would note that the quantitative immensities of the stellar world as proclaimed by contemporary astrophysics, be they factual or not, raise the question of their human assimilation. The issue, moreover, proves to be vital, for it determines whether in the end these proposed immensities will serve to guide and enlighten or to blind us, and blight our humanity. I contend that only a true metaphysics—a metaphysics that is profoundly theological—can save us from the latter outcome. "*The heavens declare the glory of God, and the firmament sheweth his handiwork*": only on that basis, I say, can we bear the immensities of the stellar world. It behooves us to realize that the heavens we perceive, be it directly or with the aid of telescopes, exemplify the heavens which we do *not* perceive, and that the quantitative vastness of the stellar universe mirrors the true immensity of the spiritual world. It is not a question here of symbolism in the anemic sense of "the merely symbolic," but in the

Platonist sense, rather, of actual "participation." The stellar universe, namely, "participates" in the authentically spiritual world, and it is this *ontological* fact that bestows a superior dignity upon the stars, a *sacredness* one can almost say, which man is obligated to respect.

To the astrophysicist, on the other hand, a star is simply "a very hot gas," and nothing more. I have argued that such a reduction is epistemologically unfounded and metaphysically untenable; it remains to comment on its effect upon humanity. From a traditional point of vantage, to be sure, the astrophysical reduction is a profanation, a kind of sacrilege; but what impact does it have upon an already profane civilization? Does a cosmic symbolism retain any kind of efficacy when it is no longer recognized, no longer understood? I surmise that the efficacy of an authentic symbol survives its comprehension: symbols do not die. The stellar universe, I maintain, retains a paramount iconic significance even in the present iconoclastic age: it is only that its significance has become inverted (here it is again: the previously mentioned *diabolic* connection!). What we think about the stars, how we picture the stellar world, does still have its effect on us; whether we realize it or not, it does influence and profoundly affect our views regarding God, man, and human destiny. The heavens, I contend, will declare either "the glory of God" or the supreme futility of existence: here there can be no middle ground, precisely because the stellar world, in its iconic function, signifies the highest cosmic sphere. If that sphere consists simply of particles engaged in meaningless motion, then all human aspirations must in the end prove vain. If the stellar light, which the ancients thought to be of celestial origin, and Plato viewed as the carrier of intelligible essences—if that light fails, the cosmos and all that it contains is reduced in the final count to nothingness. It is surely no accident that the rise of astrophysics has been accompanied by the advent of postmodernist nihilism in its philosophic as well as its cultural manifestations. The drift into nihilism corresponds precisely to the loss of substance implicit in the physicist's worldview: culture and cosmology, it turns out, are intimately linked. In fact, *as the cosmology flattens, so invariably does the culture.*

To conclude: the hierarchic distinction between stellar and terrestrial substances is vital to a sound cosmology. What I have previously termed the rediscovery of the corporeal needs therefore to be followed by another basic recognition: *the rediscovery,* namely, *of the stellar world.*

7

The Status of Geocentrism

If there has been little debate in recent times on the question of geo-centrism, the reason is obvious: it is taken for granted that the geo-centrist claim is a dead issue, on a par with the flat-Earth hypothesis. Admittedly, the doctrine has yet a few devoted advocates in Europe and America, who are neither naïve nor uninformed; yet hardly any-one cares, hardly anyone is listening. Even the biblically oriented creation-science movement, which of late has gained a certain pres-tige and influence, has for the most part disavowed geocentrism. Yet the fact remains that geocentrist cosmology constitutes not only an ancient, but indeed a metaphysically-based doctrine, which as such deserves serious consideration, to say the least. To maintain, more-over, that this cosmology has nothing to say on a cosmographic plane—that it is "merely symbolic" in other words—is to join the tribe of those who would "demythologize" just about any traditional doctrine or belief at the behest of the scientific establishment. It will not be without interest, therefore, to investigate whether the geocen-trist claim—yes, understood *cosmographically*—has indeed been ruled out of court. I shall argue that indeed it has not. As regards the Galileo controversy, I propose to show that Galilean heliocentrism has proved to be untenable, and that the palm of victory belongs in fact to the wise and saintly Cardinal Bellarmine. I should add that the problematic of this article will lead us, in the final section, to elicit an interpretation of relativistic physics which accords with tra-ditional doctrine.

Nothing perhaps is more impenetrable to the modern mind than the ancient cosmologies. What troubles us is not so much the fact that these doctrines are inherently metaphysical, but that even so they are descriptive of the perceptible world, at least in a qualitative way. When such cosmologies speak of the Sun, Moon, and stars, the reference is no doubt symbolic, but yet not "merely symbolic," which is to say that the cosmologies in question stake also a *scientific* claim. It is this fusion of a "metaphysics" with a "physics" that baffles us the most. One needs of course to bear in mind that the kind of science at issue is worlds removed from the Baconian, both in point of method and in the end to be achieved. It appears, first of all, that ancient cosmologists did not feel obliged to account for such things as retrograde planetary motions, or the precession of equinoxes; as Thomas Kuhn points out: "Only in our own Western civilization has the explanation of such details been considered a function of cosmology. No other civilization, ancient or modern, has made a similar demand."[1] A shift from cosmology of the ancient kind to astronomy in the modern sense can moreover be discerned in Greece between the time of Pythagoras and Ptolemy. One has the impression that mythical and inherently metaphysical notions, such as that of the heavenly "spheres," were being gradually transformed into "physical" conceptions destined to be scorned and discarded with the advent of modern times. Meanwhile mathematical techniques of increasing complexity and sophistication were being devised in an ongoing effort to attain ever greater accuracy in the description of observable phenomena.

The simplest part of Greek astronomy pertains to the stellar sphere. It turns out that stellar orbits can be described, with what seemed at the time to be perfect accuracy, on the assumption that the stars are fixed on the surface of an immense sphere, concentric with the Earth, which rotates diurnally around an axis that could be identified within one degree by the position of the North Star, while the Earth remains stationary at the center of the universe. If it were not for the Sun and other "wanderers," the simple "two-sphere" model of ancient astronomy would have provided a seemingly perfect description of the relevant phenomena. The orbits of these wanderers, the so-called planets, proved however to be complex and

---

1. *The Copernican Revolution* (New York: MJF Books, 1985), p. 7.

challenging in the extreme. Eventually Greek astronomers concluded that circular orbits concentric with the terrestrial sphere will not suffice for their description, and by the time of Hipparchus, whose active life can be dated between 160 and 127 BC, the method of epicycles and deferents had come into use. The motion of a planet was now conceived as the sum of two circular motions: a small rotation, namely, around a center known as the deferent, which itself sweeps out a much larger circle centered upon the Earth. The epicycles were thus conceived as a small correction to the circular orbits of the older "spherical" astronomy. I would add that in addition to his work relating to epicycles, Hipparchus is said to have discovered the precession of the equinoxes, estimated to be 36 seconds of arc per year, which compares not too badly with its actual value of some 50 seconds. He also estimated the distance to the Moon to be 33 times the diameter of the Earth, which again compares rather well to its actual value of approximately 30.2. Following upon these discoveries, progress in planetary astronomy continued apace. It was not long before astronomers realized that orbits could be progressively corrected by adding epicycles to epicycles in an indefinite series. They discovered, moreover, that additional corrections could be introduced by displacing the center of a deferent. The resultant circles, known as eccentrics, could be adjusted to further improve the result. In addition, it was found that an even higher degree of accuracy could be achieved by taking the rate of rotation of a deferent or some other point in the geometric scheme to be uniform, not with respect to its actual center, but with respect to a displaced center known as the equant. The method of epicycles came thus to be supplemented by the use of eccentrics and equants. It appears that Greek astronomy furnishes the first example of mathematical modeling on a serious scale. The development reached its zenith in the work of Claudius Ptolemy, whose treatise known as the *Almagest* (circa AD 150) dominated Western astronomy till at least 1543, when it began to be displaced by the Copernican theory.

What motivated Copernicus to reject the Ptolemaic theory in favor of a heliocentric astronomy? In his preface to the *De Revolutionibus*, Copernicus cites persistent inaccuracy and lack of coherence as his principal criticism of the prevailing astronomy. Some fourteen hundred years after the publication of the *Almagest*, the computational problems of planetary astronomy had not yet been solved with satisfactory precision. Worse still, there seemed to be no principle, no rhyme or reason governing the proliferation of epicycles, eccentrics and equants: these mathematical parts and pieces could not be made to constitute a coherent whole. "It is as though an artist were to gather the hands, feet, head and other members for his images from diverse models" complains the Polish astronomer, "each part excellently drawn, but not related to a single body, and since they in no way match each other, the result would be monster rather than man." It has been pointed out that this perception reflects the Neoplatonist influences to which Copernicus was demonstrably exposed. What in any case renders the expanded Ptolemaic model monstrous in the eyes of Copernicus is the *ad hoc* character of its multiple constructions, which is to say, its lack of mathematical intelligibility as a whole. By way of contrast he maintains that the newly discovered heliocentric astronomy exhibits "an admirable symmetry" and "a clear bond of harmony in the motion and magnitude of the spheres." Consider, for example, the retrograde motion of planets: why should a planet reverse its normal eastward motion and retrogress for a time, till it resumes its eastward course? From the standpoint of Ptolemaic astronomy, this phenomenon presents itself as an inexplicable irregularity, which can indeed by accounted for through the introduction of appropriate epicycles, but can hardly be understood. When viewed in a heliocentric perspective, on the other hand, retrograde motion becomes instead a mathematical consequence of the fact that the Earth itself is in motion around the Sun. It is easy to see that such is the case. Simply draw three concentric circles representing the sphere of the stars, the orbit of the Earth, and the orbit of the planet. In the case of an inferior planet (Mercury or Venus namely), the planetary circle will be the innermost, whereas, for a superior planet, the orbit of the Earth will be contained within the other two. If now we mark successive positions of Earth and planet on their respective circles and connect corresponding points by a line to obtain "observed" planetary positions on the stellar sphere, we can readily see how the motion of the Earth gives rise to the phenomenon of retrogression. So too one discovers that retrogression occurs when the planet in question approaches its minimum distance to the Earth, which explains why retrogressing planets are observed to shine more brightly. Here then—and for the first time!—we encounter a scientific explanation of the fact that planets retrogress.

A second irregularity which the new theory explains beautifully is the observed variation in the periods of orbital planetary motion. Given that the Earth itself orbits around the Sun, the time it takes for a planet to return to its starting position, as observed from the Earth, is clearly not the same as the actual time it takes to complete an orbit. The observed variations in planetary orbital periods can therefore be explained, once again, by the postulated motion of the Earth. Copernicus cites many examples of this kind to document the explanatory power of the heliocentric hypothesis. I will mention one more: it is known that the planets Mercury and Venus can only be observed in the vicinity of the Sun. To account for this fact in Ptolemaic terms, it was necessary to introduce deferents and epicycles binding these planets to the Sun. Heliocentric astronomy, on the other hand, requires no such *ad hoc* constructs: the given phenomenon is an immediate consequence of the fact that the orbits of Mercury and Venus are contained *within* the orbit of the Earth.

This brings us to a major point of difference between Ptolemaic and Copernican astronomy. In the heliocentric scheme, the order of the planetary orbits can be determined from observational data. If the planets (including the Earth) revolve in circular orbits around the Sun, it is possible in fact to calculate the ratios of the planetary radii in terms of the angular distances from the Sun to the planets as measured from the Earth. Such is not the case in a geocentric system, where not even the order of the planets can be determined. In a word, the new astronomy is far more coherent than the old. It is this newly-discovered coherence that Copernicus is alluding to when he speaks of "a clear bond of harmony in the motion and magnitude of the spheres"; and it is evident that he views this new "bond" as a powerful argument for the truth of his theory.

In actuality, however, the charge of incoherence applies to the new astronomy as well. In practice Copernicus was forced to introduce epicycles and eccentrics of his own, and like Ptolemy himself, ended up with over thirty circles, without any appreciable gain in the degree of accuracy. Clearly, the problem of planetary astronomy had not yet found its solution.

It appears that the decades following publication of the *De Revolutionibus* witnessed few converts. To the general public the notion of an orbiting Earth appeared both absurd and impious, and even astronomers seem for the most part to have been wary of that hypothesis. The second half of the sixteenth century was moreover dominated by the imposing figure of Tycho Brahe, a formidable opponent of heliocentrism. Brahe is known, first of all, for the uncanny accuracy of his astronomical measurements. His results are frequently precise to one minute of arc, an unrivaled achievement for naked-eye observation. What especially concerns us, however, is the fact that Brahe proposed a remarkable planetary astronomy of his own, which to this day finds partisans in Europe and America.[2] Accepting the traditional notion of an immobile Earth and a stellar sphere engaged in diurnal geocentric rotation, he proposed that Mercury, Venus, Mars, Jupiter, and Saturn circle the Sun, while the Sun and Moon circle the Earth. It happens that this geocentric theory embodies all the advantages previously cited by Copernicus in behalf of his heliocentric model. The Tychonian theory no less than the Copernican explains such things as retrograde motion, the variation of planetary periods, and the binding of inferior planets to the Sun, without recourse to epicycles or other *ad hoc* constructions. In point of fact, it can be shown that the two theories are mathematically equivalent, which is to say that they lead to exactly the same apparent planetary trajectories.[3]

---

2. It may be safe to say that all serious partisans of geocentrism today are Tychonians.

3. A sketch of the proof can be found in Thomas Kuhn, *The Copernican Revolution*, op. cit.

It should however be noted that the two theories are not equivalent in regard to stellar astronomy, for it is evident that a displacement of the Earth would entail a corresponding parallactic shift in the apparent position of a star. Tycho Brahe himself had searched for such a shift, but found none. This means that stellar parallax, if it exists, must be of an order of magnitude less than a minute of arc, which would necessitate stellar distances far greater than astronomers were wont to assume. Copernicus himself had recognized that his hypothesis demands an enormous enlargement of the stellar sphere, and it may be worth noting that Tycho Brahe considered this fantastic multiplication of apparently empty space to be one of the most cogent reasons for rejecting the Copernican hypothesis. Yet, from a strictly scientific point of view, a decision between the two theories could not be made at the time.

Although the problem of planetary astronomy, as I have said, had not yet been solved, it turns out that both Copernicus and Tycho Brahe, each in his own way, had made decisive contributions which were soon to lead to a definitive solution. The breakthrough came in the first decade of the seventeenth century at the hands of Johannes Kepler. Availing himself of superior data supplied by Tycho Brahe, he proposed a new heliocentric theory which was destined to carry the field. After years of futile endeavor, Kepler abandoned the time-honored method of epicycles in favor of a radically new idea: he proposed that the planets revolve around the Sun in elliptical orbits, and with variable speed. His so-called First Law stipulates that the Sun is situated at one of the two foci of the planetary ellipse, while his Second Law states that the line segment from the planet to the Sun sweeps out equal areas within the ellipse in equal times. Qualitatively, this simply affirms that the planet moves faster the nearer it is to the Sun; in point of fact, however, Kepler's "law of equal areas" enables one to calculate the velocity of the planet at every position of its trajectory. In conjunction, the two laws lend themselves to a complete description of the planetary system. No need any longer for epicycles, eccentrics, equants, or other devices of the kind: it turns out that two simple and mathematically elegant laws suffice to solve the age-old problem. As Thomas Kuhn points out: "For the first time a single uncompounded geometric

curve and a single speed law are sufficient for predictions of planetary position, and for the first time the predictions are as accurate as
the observations."[4] I will note in conclusion that Kepler first
recorded his new ideas in a treatise on the motion of Mars, the most
challenging of the planets: it could well be said that the era of modern astronomy commences with the publication of this work, in the
year 1609.[5]

It was in the same year, exactly, that Galileo Galilei first turned
his newly-invented telescope to the sky with startling results. In
quick succession he discovered the Milky Way to be a sea of stars,
detected mountains and craters on the Moon, the height and depth
of which he could estimate from shadows, found dark spots on the
Sun, showing that the Sun itself rotates around its own axis, and
discovered that Jupiter has four moons. While none of these findings have a direct bearing on the Copernican issue, they have had a
decisive impact upon the European mentality in that they appeared
to discredit the categorical distinction between the celestial orbs,
which from times immemorial mankind had taken to be perfect
and immutable, and the "sublunary" world: this imperfect and
ever-changing domain that constitutes our habitat. Peering through
his telescope, Galileo seemed to behold one and the same kind of
world wherever he looked: from the rugged landscape of the Moon
to the moving spots on the Sun itself. It is hard for us to imagine the
excitement in European society stirred by reports of these new vistas. A widespread fascination with astronomical discoveries seems
to have ensued, eliciting varied reactions. John Donne—to cite perhaps the most striking example—appears to have sensed the deeper
significance of the Galilean "movement" almost instantly: "And new
philosophy calls all in doubt," he penned back in 1611; "'Tis all in
pieces, all coherence gone." For the most part, to be sure, the
response was less prophetic; as Kuhn points out on the lighter side:
"The telescope became a popular toy." Yet, at the same time, far

4. Ibid., p. 212.
5. It would take us too far afield to comment on the Neoplatonist influences
that shaped Kepler's scientific convictions, a matter we shall have occasion to touch
upon in the next Chapter.

more than a toy! There can be no doubt that the new images, gleaned through that "toy," have contributed decisively to the demise of the ancient *Weltanschauung*.

One more Galilean discovery needs to be mentioned: the phases of Venus, namely, which seemed indeed to imply that Venus orbits around the Sun. But whereas Galileo exhibited this discovery as proof of the Copernican hypothesis, the fact remains that the phases of Venus are accounted for equally well on the basis of Tychonian astronomy. "It was not proof," writes Kuhn, "but propaganda."

Where, then, did the Copernican issue actually stand at the time of the Galileo controversy? One sees in retrospect that it stood very much as Cardinal Bellarmine had stated in his letter to Foscarini, in 1615: "To demonstrate that the appearances are saved by assuming the Sun at the center and the Earth in the heavens is not the same thing as to demonstrate that in fact the Sun is in the center and the Earth in the heavens," writes the Cardinal. "I believe the first demonstration may exist," he goes on, "but I have very grave doubts about the second. . . ." Yes: the discoveries of Johannes Kepler do indeed clinch the first demonstration alluded to; but as regards the second, the subsequent history of science has fully justified the Cardinal's "grave doubts." As I propose to show, the physics of our day has in fact rendered the second demonstration unthinkable.

With the publication of Newton's *Principia* in the year 1687 the triumph of Keplerian astronomy was complete. Kepler's First and Second Laws could now be derived theoretically by means of a brilliant new physics, a science that could be tested and verified in a thousand ways. So far as the scientists and the scientifically educated public were concerned, geocentrism was now a dead issue. No one doubted any longer that the Earth does move; it only remained to carry out experiments that could detect and measure that motion. What were these experiments, and what did they prove?

One approach was based upon the phenomenon of aberration. In 1676 a Danish astronomer named Olaus Roemer noted that the period between observed eclipses of one of Jupiter's moons varies by several minutes, depending upon the relative position of the Earth. He concluded that light propagates at a finite velocity, which he estimated to be 309,000 kilometers per second; a result, one might add, which is accurate to within 3%. Now, if the Earth itself moves, that additional velocity will cause a shift in the apparent position of a celestial object. Think of a car driving through rain on a windless day. Relative to the car, the rain falls, not vertically, but at an angle, which depends upon the ratio of two velocities: the horizontal velocity of the car, namely, divided by the vertical velocity of the rain. From a measurement, therefore, of the so-called angle of aberration, one can determine the ratio of the velocities in question. That is the idea behind what appears to be the first experiment designed to demonstrate and measure the orbital velocity of the Earth.

In 1724, James Bradley, the British Astronomer Royal, attached a telescope to the top of a chimney and began to observe the star Gamma Draconis, situated almost ninety degrees above the horizon. As expected, he found that in the course of a year the apparent position of the star described a small circle, corresponding to an angle of aberration close to 20 seconds of arc. By simple trigonometry one knows that this angle equals the arctangent of v/c, where v denotes the orbital velocity of the Earth and c the speed of light. It follows that an aberration of 20 seconds corresponds to an orbital velocity close to 30 kilometers per second, in good agreement with astronomical theory *à la* Kepler and Newton. Bradley had apparently proved that the Earth does move: Galileo's celebrated *Eppur Si Muove*, so it seemed, had at last been confirmed, vindicated before the world.

The story, however, does not end at that point. In 1871 another British astronomer, named George Biddell Airy, conducted an experiment based upon an idea proposed more than a century earlier by a Jesuit named Boscovich. It occurred to Boscovich that if Bradley's telescope had been filled with water in place of air, the resultant angle of aberration would have been increased, due to the fact that the velocity of light is less in water than in air. However, when the revised experiment was finally carried out, it turned out that the angle in question had not changed at all! The claim that Bradley's shift of 20 seconds is caused by aberration had thus been disproved; and to everyone's amazement, the argument *against* the motion of the Earth seemed now to be compelling. The logic is simple: given that orbital motion implies aberration, it follows that *no aberration* implies *no* orbital motion. And needless to say, this recognition sent shock waves through the scientific community.

However, worse was yet to come. In 1887 Michelson and Morley conducted their famous experiment designed to detect and measure the orbital velocity of the Earth, not by way of aberration, but by comparing the observed velocity of light in the direction of that orbital motion with its velocity in the opposite direction. Evidently the two velocities should differ by exactly 2v, where v again denotes the orbital velocity. But once again to everyone's consternation, the two light velocities turned out to be exactly the same, implying that v equals *zero*. Two crucial experiments, based upon different physical principles, had now led to the same conclusion: *the Earth does not in fact move.*

At this juncture one has only two options: one can accept the verdict that the Earth does not move and opt for a duly refined Tychonian astronomy, or search for a way of adjusting the laws of physics so as to render the Earth's orbital velocity to be in fact undetectable. History records that Albert Einstein opted for the second alternative, and in so doing astonished the world with his theories of relativity. The special theory in particular, published eighteen years after the fateful year 1887, ingeniously "explains" the negative results of both the Airy and the Michelson-Morley experiments. Regarding the scientific credentials of Einsteinian physics, suffice it to say that although neither the special nor the general theory may in fact be quite as well-founded as one has been led to believe, they constitute doubtless one of the most brilliant and successful ventures in the history of modern science.[6] However, one must not forget that Einsteinian physics operates by the logic of "saving appearances," which

---

6. An impressive body of *contrary* evidence has been extracted from the scientific literature by Robert J. Bennett and Robert A. Sungenis in their monumental 2-volume treatise entitled *Galileo Was Wrong*, the fifth edition of which appeared in

is not the same thing, to paraphrase Cardinal Bellarmine, as demonstrating that what is claimed is in fact the case. As Walter van der Kamp—that indefatigable champion of Tychonian astronomy—was fond of pointing out, the logic of relativity constitutes a *ponendo ponens* argument: from the premise "P implies Q" one falsely concludes "If Q, then P."

Getting back to Bradley's experiment, the question remains: if indeed the Earth does *not* move, and there is consequently *no* stellar aberration, how then is the small circle described by Gamma Draconis to be explained? On a Ptolemaic or Tychonian basis the answer is clear: the observed phenomenon must be caused by an actual circular motion of Gamma Draconis relative to the stellatum, the revolving sphere of the stars. A similar remark applies to the phenomenon of stellar parallax, which was in fact finally detected by Henderson in 1832, and accurately measured in 1838 by Bessel and Struve. Now, from a geocentric point of view there obviously *is* no such thing as stellar parallax, which is to say that the observed phenomenon must be caused, once again, by corresponding stellar motions. The fact, however, that such motion has not been observed by no means implies the existence of stellar parallax.

This is not to say, however, that the geocentrist issue is a matter of indifference to stellar astronomy: nothing could be further from the truth. The fact is that the hypothesis of parallax plays a key role in the present-day determination of stellar distance. This can already be seen from the circumstance that the standard unit of astronomic distance is the so-called parsec, the distance, namely, at which a baseline, whose length equals the mean distance from the center of the Earth to the center of the Sun, subtends an angle of 1 second. Clearly, the parsec is a parallactic unit of length: it says that a star situated 1 parsec from the Earth will have parallax on the order of 1 second. The fact, therefore, that the parsec turns out to be approximately 3.26 light-years is indicative of the stupendous distances the

---

2008. Despite occasional deviations from the standards of academic discourse on the part of the principal author, the work constitutes an invaluable critique of the contemporary scientific status quo as it pertains to the Galileo issue. It is to date perhaps the definitive text on that subject.

hypothesis of stellar parallax imposes upon the universe. Moreover, since most stars—all but a relative few—have no measurable parallax at all, their distance from the Earth must be at least 25 to 50 parsecs. And as if this were not distance enough, stellar astronomers are wont nowadays to employ the megaparsec as their preferred unit! A Tychonian universe, by comparison, would still need considerable size: but nothing like the billions of light years to which contemporary astronomy lays claim. One sees that a geocentric interpretation of astronomical phenomena, by eliminating stellar parallax, would substantially undercut our current notions regarding the stellar world: a Tychonian universe would, in particular, differ radically from the big bang model in size as well as in architecture.

But whereas contemporary astronomy is thus implacably opposed to the geocentrist hypothesis, it happens that pure physics is not. According to general relativity, it is in fact permissible to regard the Earth as a body at rest: as Fred Hoyle has put it, the resultant theory "is as good as any other, but not better." Relativity implies that the hypothesis of a static Earth is not incompatible with the laws of physics and cannot be experimentally disproved. Of course physics as such cannot affirm that hypothesis; *but neither can it deny its validity.* Already in 1904 Henri Poincaré had noted that "the laws of physical phenomena are such that we do not have and cannot have any means of discovering whether or not we are carried along in a uniform motion of translation";[7] and by 1915 Einstein had concluded that the same applies to arbitrary reference frames. Thus, so far as physics is concerned, the geocentrist claim remains viable.

Now, given that the scientific challenge to geocentrism derives, not from physics, but from astronomy, we need to ask ourselves whether the latter is in a position to prove its case. Our scientific knowledge concerning the stellar universe is of course based upon observations carried out either in terrestrial observatories or by means of instruments transported into outer space. It is crucial to note that the transition from observational data to claims concerning the stellar realms cannot be accomplished on the ground of

---

7. Quoted by Dean Turner in *The Einstein Model and the Ives Papers* (Greenwich: Devin Adair, 1979), p.154.

physics alone, but requires additional hypotheses of an untestable kind. We have already encountered one such hypothesis in the assumption of stellar parallax, and the reader may recall the Doppler interpretation of stellar redshifts, along with the so-called Copernican principle, as additional assumptions needed to underpin the contemporary astronomy.[8] The logic here is once again of the *ponendo ponens* variety, which is to say that the hypotheses in question are judged or validated by their apparent success in "explaining" observable phenomena. But apart from the circumstance that, as Cardinal Bellarmine has pointed out long ago, such an argument is never compelling, it happens that the prevailing astrophysical cosmology has so far failed the empirical test: despite inflated reports and decades of concerted effort, one finds that big bang theory does not actually square with the empirical data taken in its entirety. If the Airy or Michelson-Morley experiments had yielded their intended result, the scientific case against geocentrism, though still not compelling, would have been at least impressive; as the matter stands, however, the ancient doctrine has not even been rendered implausible, let alone disqualified.

The late Walter van der Kamp has made an interesting point: "In truth," he said, "the choice is between Tycho Brahe and Einstein—Galileo, *et al.,* are played out." Now, I certainly agree with the part about Galileo. To be precise, he is "played out," not because he affirmed the motion of the Earth, but because he staked the claim on ostensibly scientific grounds. One knows today, on the basis of physics itself, that Galileo's arguments are inconclusive, and that in fact the stipulated motion cannot be proved at all. What, on the other hand, I find misleading in van der Kamp's assessment is that the choice between Tycho Brahe and Einstein is presented as an "either/or" alternative. In other words, I reject the implication that Tychonian astronomy and Einsteinian physics are mutually exclusive. Not only shall I argue that the two positions are logically com-

---

8. We have touched upon both issues in Chapter 6.

patible, but that each may in fact have its own validity. The key to the problem, once again, lies in the discernment of what may be termed "levels of cosmic reality": though interrelated, these remain distinct and need to be distinguished.

To begin with the physical, let me recall that I use this term in a technical sense: it refers to the level or aspect of cosmic reality that answers to the *modus operandi* of physical science. The physical is thus defined and known through acts of measurement, which entails that it owns neither essence nor substance in the traditional ontologic sense. As Heisenberg observed with reference to quantum particles, physical objects constitute indeed a strange new kind of entity "just in the middle between possibility and reality."[9] Now, I will argue that on the level of these "strange new entities," Einsteinian relativity has its place. It is by no means surprising, I say, that entities defined through acts of measurement should conform to principles of relativity: in the final analysis, are they not in fact *relational* by definition? Consider the case of velocity, the magnitude of motion: what else can it be, physically speaking, than a speed relative to this or that coordinate system, this or that frame of reference? On the physical plane there are indeed *invariants*—quantities, namely, which turn out to be the same relative to every coordinate system in some class—but there are no absolutes: there is nothing physical which is not inherently relative. More than that requires *essence*, something that answers to the question "What?". A stone thus—or a cat!—is not relative, not simply relational, by virtue of the fact that it *is* something. It is something *more*, thus, than "midway between possibility and reality." A stone or a cat, therefore, cannot be quantified, cannot be elicited by acts of measurement, *and consequently does not pertain to the physical domain*. As Eddington keenly observed: "The concept of substance has disappeared from fundamental physics."[10] Strictly speaking, physics does not deal with *things*. It is true that the notion of substance was retained in what is now called classical physics; that retention, however, proved

9. *Physics and Philosophy* (New York: Harper & Row, 1962), p. 41.
10. *The Philosophy of Physical Science* (Cambridge University Press, 1949), p. 110.

to be spurious, inconsistent with the operational principles of physical science, which is of course why it was eventually cast out. As Eddington points out: "Relativity theory made the first serious attempt to insist on dealing with the facts themselves. Previously scientists professed profound respect for the 'hard facts of observation'; but it had not occurred to them to ascertain what they were."[11] And this means that the tenets of relativity are not merely *ad hoc* stipulations introduced to safeguard the Copernican hypothesis as geocentrists are wont to claim, but respond to the operational principles upon which modern physical science is based. One is beginning to see that in a world defined operationally—in what John Wheeler terms the participatory universe—Einsteinian relativity does reign supreme.

But how does the matter stand on the level of the corporeal world? Do the principles of relativity still apply to a world that is to be known, not by way of measurement, but through acts of cognitive sense perception? Do Einsteinian postulates hold in a world where essences manifest in the form of sensible qualities, a world in which not only *relations* but *substances* are to be found? There is actually no reason to believe that such is the case. Nothing obliges us to suppose that in this corporeal domain, which is distinctly "more" than the physical, there can be no absolute rest and absolute motion, nor absolute simultaneity of events.[12] If, on this higher level, stones and cats are real, why not also rest and motion, and why not simultaneity of events as well? I have argued in *The Quantum Enigma* that it is the loss of essence, of substantial being as one descends to the physical plane, that brings into play the quantum-mechanical superposition principle; I would like now to propose that the same loss, the same reduction, may lead macroscopically to

11. Ibid., p.32.
12. One is reminded of René Guénon's remarkable claim to the effect that those who cannot conceive of all events as occurring simultaneously are debarred from even the least understanding of metaphysics. This obviously places the metaphysical realm at the very antipode of the physical, in which no two distinct events can be absolutely simultaneous. It appears that as one ascends the ontological *scala naturae*, temporal dispersion is progressively diminished and eventually transcended.

Einsteinian relativity. *Where there is no substance to distinguish one reference frame from another, it is hardly surprising that the two are of equal weight.* It is, as always, the loss of substance, of hierarchy in fact, that leads to the "democratization" of what remains.

One knows that the superposition principle is abrogated on the corporeal plane: one knows, for example, that a cat cannot be both dead and alive. It is this simple fact that accounts for the so-called collapse of the state vector, which has mystified physicists since the advent of quantum theory in 1926.[13] Is it not reasonable, then, to suppose that the principles of relativity are likewise abrogated on the corporeal plane? Now, this abrogation, which I take to be factual, has obviously immense implications. It suggests, for example, that Aristotelian physics may not, after all, be quite as chimerical as we generally assume. What presently concerns us, however, is that it casts evidently a new light on the geocentrist controversy. One sees, in particular, that Tychonian astronomy may be more than a merely "admissible" theory, as general relativity declares: more indeed than simply "as good as any other, but not better." The point is that the worth of geocentric astronomy can no longer be ascertained exclusively through acts of measurement: the question, one now finds, cannot be resolved on purely physical grounds, but calls for considerations of a different order. *The most that physics as such can say is that geocentric astronomy cannot be ruled out of court.*

To summarize: Geocentrism is the cosmology at which one arrives by way of cognitive sense perception, whereas Einsteinian acentrism answers to the cognitive means of the contemporary physical sciences. There can in truth be no conflict, no contradiction between the two: the respective worldviews correspond simply to different perspectives, different *darshanas* as the Hindus would say. However, geocentrism is the higher of the two, even as the corporeal plane is above the physical ontologically. Cognitive sense perception, moreover, having access to essence, is able in principle to transcend the corporeal plane: to pass, in the words of St. Paul, from *"the things that are made"* to the *"invisible things of God"*— and beyond even these, to *"His eternal power and Godhead."* In a

13. See Chapter 1.

word, whereas human perception opens in principle to the meta-cosmic realms, the *modus operandi* of physical science confines us to a relational and indeed subcorporeal domain. Galilean heliocentrism, finally, constitutes a bastard notion which spuriously confounds the two ways of knowing. One might add that there is also a traditional or authentic heliocentrism, which must not be confused with the Galilean; and this is what will mainly concern us in the following chapter.

8

Esoterism and Cosmology:
From Ptolemy to Dante and Cusanus

There are doctrinal conflicts which can only be resolved on an esoteric plane. In the present chapter, I propose to reflect upon one
such discrepancy: the antithesis, namely, between a geocentric and a
heliocentric worldview. It happens, however, that there is more than
one geocentrism, even as there are several distinct kinds of heliocentrism. It is necessary, therefore, to sort out these various conceptions, which pertain to different levels and must not be confounded:
only then can we grasp the crux of the problem.

In the first place it is needful, once again, to distinguish between
two very different ways of knowing: the way of cognitive sense perception, which takes us into the corporeal domain, and the *modus
operandi* of physical science that gives access to what I term the
physical universe. This said, it becomes apparent that the primary
geocentrism—the geocentrism natural to mankind—is based upon
the first way of knowing: looking up at the night sky, one actually
perceives the stars and planets circling the Earth, while the latter
itself is experienced as central and immobile. As regards the second
way of knowing, one generally takes it for granted that physical science has come down unequivocally on the side of heliocentrism.
But as we have come to see in the preceding chapter, it happens that
contemporary physics does allow the hypothesis that the Earth does
not move, does not in fact orbit: according to Einsteinian theory, no
experiment can possibly prove otherwise. Admittedly, this is not
much of a geocentrism; but so far as scientific knowing is concerned, it is the most that can be said: *physical* geocentrism let us
call it, to distinguish the latter from the primary kind. To be sure,

there is also a physical heliocentrism which affirms that it is likewise admissible to consider the Sun to be at rest, and conceive of the Earth as orbiting around the Sun. On the level of physical theory there is no conflict between the two positions, which is to say that both are validated by the principle of relativity. As has been suggested in the preceding chapter, that principle is expressive of the fact that the notion of substance has no more place in fundamental physics: in a world consisting as it were of relations, Einsteinian relativity does indeed reign supreme.

It should be noted that there exists evidently no heliocentrism based upon cognitive sense perception. Yet in addition to what I have termed *physical* heliocentrism there is the position championed by Galileo, which insists on supposedly scientific grounds that the Earth does revolve around the Sun. As we have just seen however, Galileo's arguments prove to be fallacious, and his celebrated *"eppur si muove"* turns out to be in fact unprovable. What I shall term *Galilean* heliocentrism—the imperious claim which to this day continues to define our collective *Weltanschauung!*—proves thus to be finally no more than a spurious hybrid of the two aforesaid ways of knowing.

There is also, however, a third kind of heliocentrism, which might be termed *traditional, iconic*, and even perhaps *esoteric*; we will consider that heliocentrism in due course. But first it behooves us to reflect in some depth on the meaning and significance of the primary geocentrism.

It has been suggested that the geocentrist worldview corresponds to the mentality of so-called primitive man, someone who accepts the testimony of the senses uncritically and is supposedly incapable of scientific thought. One maintains moreover that human perception is inherently unreliable and subject to manifold illusions, and that these need to be rectified through scientific means before authentic knowledge can be attained. Even scientists admit, of course, that sense perception does indeed constitute our one and only means of access to the external world; but one denies that it can *per se* bestow authentic knowledge of things as they are. For that one needs to supplement the human faculties by scientific instruments, and avail oneself of the theories which underlie their use. The role of sense perception in the cognitive process is thus ultimately reduced to elementary acts, such as the reading of a pointer on a scale.

Oversimplified as this brief characterization of the science-oriented epistemology may be, it does serve to identify the contemporary scientistic denigration of sense perception as a serious and respectable way of knowing. To the scientistic mentality the *modus operandi* of science ranks in the final count as the sole means by which authentic knowledge can be obtained; as Bertrand Russell has famously put it: "What science cannot tell us, mankind cannot know." But of course this is by no means the case! We need to understand from the outset that cognitive sense perception can in fact give access to domains of reality beyond the range of scientific inquiry, and that what we apprehend in our daily life is part and parcel of an authentic world which physical science as such cannot reveal. We need further to understand that cognitive perception is neither a physiological nor indeed a psychological act, but is in fact consummated in the authentic *intellect*, the highest faculty within the human compound: so high in fact, that according to Platonist philosophy it transcends the very bounds of space and time. Thus, even in its humblest quotidian manifestations, cognitive sense perception proves to be something quite miraculous, something literally "not of this world."[1] What actually limits the truth and the depth of human perception, moreover, are not our faculties as such, but the use we make of them; and in this regard it appears that a collective decline has been in progress since primordial times. It appears, moreover, that the scientistic denigration has itself had a debilitating effect upon our capacity to perceive, and has in fact accelerated our collective descent from the pristine state in which, according to St. Paul, man could penetrate "*the things that are made*" so as to apprehend "*the invisible things of God*" which they exemplify.[2]

---

1. On this subject I refer to my chapter on "The Enigma of Visual Perception" in *Science and Myth*, op. cit.

2. Rom. 1:20–22.

Getting back to the question of geocentrism, it is to be noted that the worldview at which one arrives by way of sense perception is of course geocentric. In light of the preceding reflections however, so far from constituting some kind of stigma, this fact in itself bestows legitimacy and indeed a primacy upon the geocentric *Weltanschauung*. It appears that the latter answers to the normal human outlook, which as such cannot be illegitimate or void of truth. And what we learn by way of our sensory faculties is that the Earth we stand upon reposes at the center of the universe, and that the Sun, Moon, planets and stars revolve around the Earth. To be sure, the geocentrist outlook does commend itself to the understanding of simple and untutored minds, as we have been told often enough; but it happens that this worldview is congenial to the understanding of sages and saints as well.

It should be noted in this regard that hitherto unsurmised *numerical* correspondences between the Gestalt aspects of planetary astronomy and the subtle anatomy of man have been discovered not too long ago by a German phenomenologist named Oskar Marcel Hinze which by no stretch of the imagination could be conceived as "accidental." Inasmuch as these congruities—amounting to a veritable isomorphism—disappear in a non-geocentric astronomy, one sees that geocentrism, so far from being illegitimate, has actually a certain primacy. And I would note that this statement is made from a strictly scientific point of view: this is by no means superstition or "pie in the sky"! As a matter of fact, as I have noted elsewhere, not only does Hinze's claim satisfy the criteria of scientific validity, but it is actually possible on that basis to prove by way of so-called ID theory that the current naturalistic explanation of how our solar system originated proves to be inadequate.[3]

The traditional doctrine of geocentrism is based upon the concep-

---

3. I have dealt with Hinze's discovery and its implications at considerable length in Chapter 6 of *Science and Myth*, op. cit. The subject of ID theory will be treated in the next chapter.

tion of the *stellatum*, the sphere of the stars, which rotates diurnally around the Earth. Between that celestial sphere and the Earth there are the planets, the "wanderers," which differ sharply from the stars by the complexity of their apparent motions. What is of primary significance, however, is the underlying two-sphere architecture of the cosmos: the notion of an outermost sphere comprised of stars, revolving perpetually about the spherical Earth which rests immobile at the absolute center of the universe. It is crucial to note that the distinction between the two spheres, so far from being simply cosmographical, is primarily ontological, which is to say that the respective spheres represent two distinct ontologic domains, two different worlds if you will; and it is worth noting that to this day one does speak of "spheres" in a distinctly ontological sense. So too it is crucial to understand that the two worlds—the stellar and the terrestrial—define a hierarchic order: that the stellar, namely, is *higher* than the terrestrial; and again I would point out that the adjectives "high" and "low" have to this day retained their hierarchic connotation. One sees that the two-sphere conception of the cosmos defines a dimension of "verticality" which is at once cosmographic, ontologic and axiological. The astronomical distance, be it measured in meters or in parsecs, separating the Earth from the stellatum becomes thus indicative of the immeasurable hiatus, again both ontologic and axiological, separating the two domains. One might add that the stellar world, though it cannot be identified with the spiritual which is metacosmic and invisible to mortal gaze, is yet reflective of the spiritual to a pre-eminent degree.

These indications, sparse though they be, may perhaps suffice to provide an initial glimpse of what geocentric cosmology is about. One sees, first of all, that with his telescope and his polemics Galileo has assaulted far more than a mere cosmography: it was not simply a question of whether the Earth does or does not move! Nor was the point at issue whether or not the Galilean claim contradicts certain passages in Scripture, for instance those that speak of the Sun as "rising," or as "running its course." What stands or falls is finally nothing less than an entire *Weltanschauung*. What has actually come under attack is in fact the notion of cosmic hierarchy, of "verticality" in the traditional "ontologic and axiological" sense. But let us note that inasmuch as this notion is basic to the very conception of spiritual ascent, what is implicated is finally the entire economy of the Christian life. One may object to this assessment on the grounds that it is surely possible to "ascend" spiritually without flying up into the sky; but whereas the spiritual or metaphysical sense of verticality needs indeed to be distinguished from the cosmographic, it yet remains that the two are profoundly linked: it is not mere imagination or pious poetry that Christ—and before Him, Enoch and Elias—*"was taken up, and a cloud received him out of their sight."*[4] The question remains, moreover, whether the two senses of "verticality" can in fact be separated on an existential plane, and whether the cosmographic sense may not in fact play a vital role in the spiritual life. One wonders whether an individual who thinks, *à la* Einstein, that "one coordinate system is as good as another" can in fact maintain a living belief in the operative truths of Christianity. What counts spiritually is what we believe with our entire being: inclusive, one is tempted to say, of the body itself, the corporeal component of our nature. Does not the First Commandment exhort us to love God *"with all thine heart, with all thy soul, and with all thy might"*? There can be little doubt that the ternary *heart-soul-might* corresponds indeed to the Pauline *pneuma-psyche-soma*, which is to say that we are enjoined to love God not only with our spiritual and mental faculties, but with our *corporeal being* as well. Moreover, in line with this basic principle, the Church has decreed that the literal or "corporeal" sense of Scripture cannot be denied, cannot be simply jettisoned as contemporary theologians are wont to do. Authentic Christianity has always rejected every kind of angelism; if man is indeed a trichotomous being, his religious convictions and discipline need be in a sense trichotomous as well.

Getting back to the concept of verticality, it follows then that the cosmographic sense of that notion cannot be cast aside with impunity; and I would add that history appears to bear this out: it is surely no accident that in the wake of the Copernican Revolution religious faith has visibly waned. In the more educated strata of society, especially, belief in the teachings of Christianity—to the

4. Acts 1:9.

extent that it has survived at all—has become strangely hollow and bereft of existential reality. There are notable exceptions, thank God; yet the overall trend is unmistakable: in a very real sense, Western man has forfeited his spiritual orientation. Having suffered the loss of cosmographic verticality, he finds himself in a flattened-out universe in which the concerns of authentic religion make little sense. Let no one say that religion or spirituality have no need of a cosmology: nothing could be further from the truth! As Oskar Milosz has wisely observed: "Unless a man's concept of the physical universe accords with reality, his spiritual life will be crippled at its roots." Yes, it is happening before our very eyes! As concerns Galileo and his famous trial, one cannot but commend the Church for rallying to the defense of a position which in truth is its very own.

One needs to understand that geocentric cosmology is inherently an iconic doctrine. It pertains thus to the traditional sciences as distinguished from the modern, which are concerned with the material and thus non-iconic aspects of cosmic reality. As Seyyed Hossein Nasr explains:

> The modern sciences also know nature, but no longer as an icon. They are able to tell us about the size, weight and shape of the icon and even the composition of the various colors of paint used in painting it, but they can tell us nothing of its meaning in reference to a reality beyond itself.[5]

That is just the point! A mountain of misunderstanding and confusion in the debate over geocentrism could have been avoided if the disputants on both sides had realized that the geocentrist claim is inherently an *iconic* truth, which as such transcends the purview of the physical sciences. Ultimately geocentrism has to do with cosmic symbolism, and thus with the mystery of essence: and that is not something that can be dealt with in positivistic terms.

Having characterized geocentrism as an iconic doctrine, it may

---

5. *The Philosophy of Seyyed Hossein Nasr* (La Salle, IL: Open Court, 2001), p.487.

be well to point out that the "symbolism" in question is not to be interpreted in some psychological sense: it is not a question of subjective, but of objective truth. Geocentrism is thus indeed a *scientific* doctrine, which however pertains, as I have noted before, not to the modern, but to the traditional sciences. And as such it demands a certain ability to "see," the capacity to actualize a superior faculty of vision, a kind that can discern the meaning of the icon as distinguished from mere "shapes and colors." The contemporary scientist, on the other hand, has been trained to do the very opposite, namely, to fix his gaze upon the outermost aspects of corporeal reality: is it any wonder that he misses the iconic sense? By means of considerable schooling one eventually becomes proficient in the task of reducing the icon to its shapes and colors: reducing the universe, that is, to its material and quantitative components. And so it comes about that the actual meaning of geocentrism escapes not only its scientific critics, but its contemporary scientific defenders as well: the contemporary debate is over the outer husk.[6]

Not only the reality, however, but the very conception of science in the traditional sense has been virtually lost. Even theologians, who *should* know better, have for the most part not a clue: if they had, they would not have busied themselves with the task of "demythologizing" the sacred texts. What then might be the cause of this deficiency, this veritable blindness? It is not a question of erudition, or even perhaps of "faith" in the religious sense; what is needed is a traditional ambience, something which in the West has disappeared centuries ago. Nasr is no doubt profoundly right when he compares the traditional sciences to "jewels which glow in the presence of the light of a living sapiential tradition and become opaque once that light disappears."[7] We need to realize that this marvelous metaphor applies not only to various recondite disciplines, such as alchemy or astrology, but to geocentrism as well, the meaning of which every-

---

6. It may surprise many readers that geocentrism still has scientific advocates. One of the best-known is Gerardus Bouw, director of the Association for Biblical Astronomy, editor of *Biblical Astronomy*, a journal dedicated to the scientific defense of geocentrism, and author of a highly interesting treatise entitled *Geocentricity*.

7. Op. cit., p. 488.

one presumes to understand. Given that cosmic realities are connected to their exemplars by way of essence, it follows that a worldview bereft of essence has no place for traditional science—be it geocentrism or any other—as well. Such a traditional science may of course survive in its outer forms, even as the shapes and colors of an icon remain in place when its meaning has been lost. Geocentrism, in particular, may thus survive in its cosmographic dimension; thus reduced, however, to its outermost sense, it turns indeed into a superstition, the mere vestige of a forgotten worldview. In terms of Professor Nasr's metaphor, geocentrism becomes thus "opaque."

Geocentric cosmology, whether conceived Ptolemaically or according to the Tychonian system,[8] affirms that the stars and the seven classical planets—Saturn, Jupiter, Mars, Sun, Venus, Mercury and Moon—are engaged in ceaseless revolution around the Earth, as if mounted on giant rotating spheres. In short, the heavens revolve while the Earth stands still: what is the *iconic* significance of that? To the ancients it meant that the stars and planets serve as principles of motion in the terrestrial sphere. Even as the Sun gives rise to the alternation of day and night, and of the seasons, and the Moon gives rise to oceanic tides and other phenomena, so it is with the stars and the five remaining planets: such was the ancient belief. Astronomy and astrology were thus inextricably linked, and could in fact be viewed as complementary aspects of a single science. Let us recall that Ptolemy has left us not only his *Almagest*—the most comprehensive and influential treatise on astronomy produced in antiquity—but the *Tetrabiblos* as well, which delves into predictive astrology no less.

Given that the celestial spheres do indeed exert an influence upon the terrestrial world, how then, let us ask, is that influence transmit-

8. According to Tychonian astronomy, the planets Saturn, Jupiter, Mars, Venus, and Mercury orbit around the Sun, while the Sun and the Moon orbit around the Earth. See p. 146.

ted to the sublunar realm? At the hands of Aristotle this question received a rather physical if not indeed mechanistic answer. Having convinced himself on philosophical grounds that there can be no such thing as empty space, and persuaded that the celestial spheres are composed of an element termed the aether, he thought that each sphere exerts a kind of mechanical force upon the next, from the stellatum down to the terrestrial. And since the latter sphere does not itself move, the result must be a "mixing of the elements," and thus the production of internal motion and change: such is at least the apparent sense of the Aristotelian theory. It appears, however, that the earlier conceptions of astrological influence had been far more refined than this apparently crude explanation. The pre-Aristotelian conceptions had been far more theological than physical, if one may put it so; we must remember that preceding civilizations had populated the heavens with gods—or angels, as we prefer to say—who presumably disposed over more spiritual means of communicating their influence to the sublunar realm. But be that as it may, the celestial spheres were in any case conceived as "active" in relation to the terrestrial, which is to say that the worldview of these early civilizations was inherently astrological.

This basic feature of ancient cosmology has of course been abandoned in the wake of the Copernican Revolution. Copernicus himself tried hard to salvage as much as he could of the old cosmology: he was by no means a revolutionary or an iconoclast. Yet, by a kind of relentless logic, his astronomical innovation did precipitate the collapse of the ancient worldview: in the minds and imagination of those who, following Copernicus, came to espouse the heliocentric cosmography, astrology became a dead issue. For now the Earth itself revolves, and presumably acts upon other planets, even as these in turn act upon the Earth. The new cosmology is thus visibly democratic: the traditional hierarchy, in which the Earth had been relegated to the lowest position, has been replaced by a planetary system in which the terrestrial globe enjoys more or less equal status with its companion planets. There is now no more "up" and "down," no more "east" and "west," "north" and "south," except of course in relation to a particular planet orbiting the Sun. Clearly, the very basis for an astrological outlook has disappeared.

According to the new cosmology, the stars and classical planets no longer exert an influence upon the Earth; or to put it more accurately, no longer exert a "higher" influence. According to contemporary physics, there is an interaction via gravitational and electromagnetic forces; and certainly in that sense the Sun, Moon and stars still affect the Earth. But it is needless to point out that the action of forces or exchange of particles admitted by the physics of our day are nothing like the "influence of the celestial spheres" as conceived in ancient lore—which is of course precisely the reason why the very idea of astrology appears to us today as a primitive and indeed exploded superstition.

Iconic truth has to do with the relation of a cosmic to a metacosmic reality. The reading of a cosmic icon contributes however an element of its own: a perspective or point of view one can say. And this entails that such an icon can be read in more than one way.

Having spoken of geocentrism as an iconic doctrine, it now behooves us to note that heliocentrism, rightly understood, is likewise iconic. Both contentions turn out to be correct, which is to say that each embodies an iconic truth; it is the perspective, the point of view, that differs. More precisely, the respective doctrines correspond in fact to different *levels* of vision. The heliocentric, inasmuch as it evidently entails a more intellectual or "inward" kind of vision, actually ranks higher than the geocentric. Whereas the latter, by virtue of its "earth-centered" perspective, perceives the cause and principle of all being in terms of its effect or influence upon the terrestrial sphere, the heliocentric is focused upon the Sun, which as the representative of Deity does by right occupy the center of the universe. As "the author not only of visibility in all visible things, but of generation and nourishment and growth" as Plato says,[9] it could not be conceived Ptolemaically as a mere planet, one among several bodies that revolve about the Earth. Given the overtly theophanic outlook of traditional heliocentrism it is hardly surpris-

9. *Republic* vi.

ing that the doctrine is closely associated with the Pythagorean and Platonist schools as distinguished from the Aristotelian. Based on the report of Philolaus, the Pythagoreans espoused a non-geocentric cosmology in which the Earth revolves around a central fire, the so-called Altar of the Universe, which apparently was not however identified with the Sun. That step was taken later by the Neoplatonists, whose cosmology thus became overtly heliocentric. Eventually, when the doctrine was revived in the Renaissance movement championed by Marsiglio Ficino, it assumed again a somewhat altered form: what Ficino instituted was almost a full-fledged religion, a kind of neo-paganism. Copernicus himself was profoundly influenced by this movement, as can be clearly seen from numerous passages in the *De Revolutionibus*. To cite but one example (from the tenth chapter of the First Book) that enables us to savor the spirit of those Renaissance times:

> In the middle of all sits the Sun enthroned. In this most beautiful temple, could we place this luminary in any better position from which he can illuminate the whole at once? He is rightly called the Lamp, the Mind, the Ruler of the Universe; Hermes Trismegistus names him the Visible God, Sophocles' Electra calls him the All-seeing. So the Sun sits upon a royal throne ruling his children the planets which circle round him.

Yet notwithstanding these panegyrics, it appears that the light of iconic truth was fast fading. A kind of earth-bound literalism, hostile to the spirit of Platonic philosophy, was beginning to manifest itself, foreboding the advent of the modern age. Neither in Marsiglio Ficino nor in Copernicus do we encounter an authentic revival of Platonist doctrine, nor can it be said that the resultant heliocentrism conforms to its traditional prototype: "Rather was it comparable," writes Titus Burckhardt, "to the dangerous popularization of an esoteric truth."[10]

It behooves us now to ponder this highly significant remark. Why should the truth of heliocentrism be termed "esoteric"? And why should its popularization be "dangerous"? Having characterized the

---

10. *Mirror of the Intellect* (State University of New York Press, 1987), p. 21.

truth of heliocentrism as "iconic," are we perhaps to conclude that "iconic" and "esoteric" are synonymous? By that token, however, authentic geocentrism would be "esoteric" as well.

I propose to give at least a partial answer to these questions. Let it be noted, first of all, that there is a *prima facie* opposition, a kind of logical contradiction, between the geocentric and the heliocentric claims. It is further to be recalled that heliocentrism is based upon an intellective vision which replaces or supersedes the sensory. The point, however, is that authentic heliocentrism—that is to say, heliocentrism understood *esoterically*—does not deny that sensory truth, but accommodates it, rather, within an enlarged and perforce hierarchic vision of reality. Vivekananda has put it well when he said: "Man does not move from error to truth, but from truth to truth: from truth that is lower to truth that is higher." Now, this recognition of lower truth, I say, constitutes indeed a mark or criterion of authentic esoterism. The higher truth is never destructive of the lower: quite to the contrary! A so-called esoterism, therefore, which undercuts the normal and in a sense God-given beliefs of mankind is perforce a fake: a dangerous counterfeit. Has not Christ Himself declared: "*I am not come to destroy, but to fulfill*"? And mark the words: "*For verily I say unto you, till heaven and earth pass, one jot or one tittle shall in no wise pass from the law, till all be fulfilled.*"[11] Admittedly Christ is speaking of the Mosaic law; yet one may surmise that His words apply likewise to the body of beliefs enshrined in the Old Testament, which most certainly includes the tenet of geocentrism. Till "*heaven and earth pass*" all these "lower truths" shall remain effective and in a way binding upon us: let no one cast them aside before "heaven and earth" *have* passed, on pain of falling into what an Upanishad calls "a greater darkness."

Getting back to the *prima facie* contradiction between the geocentrist and the heliocentrist claims, I would like now to point out that this paradox cannot be resolved on the level of our "common sense" views concerning corporeal reality. Nor indeed can it be resolved on an Aristotelian basis, let alone a Cartesian. It seems ultimately to require the high ground of a non-dualist metaphysics, be

11. Matt. 5:17–18.

it Platonist, Vedantic, or Trinitarian:[12] no lesser realism will suffice. And yes, that ground is indeed "esoteric," to say the least. There can be little doubt, moreover, that this too is the ground upon which Dante conceived his monumental vision of what might be termed the integral cosmos. In a single poetic cosmography he combined, if you will, the geocentrist and the heliocentrist cosmologies; and it is highly significant that one passes from the former to the latter precisely at the threshold of the Empyrean, which thus represents the boundary, as it were, between the two "worlds." For indeed, as one crosses that boundary the ascending spheres no longer expand, but now contract. In that supernal and indeed angelic realm, the hierarchic order of successive spheres is reversed: here to "ascend" means to approach the Center, which comprises the Altar of the Universe, that is to say, the Throne of God. The Empyrean, thus—the outermost Ptolemaic sphere—marks the point of reversal, where "*heaven and earth shall pass*," which is also the point where "*a new heaven and a new earth*" shall come to be.[13]

The question arises whether the preeminence of authentic heliocentrism may not be reflected on the physical plane in some salient cosmographical characteristic: does not the very principle of cosmic symbolism demand that the superior glory of the true heliocentric vision be mirrored somehow in the actual geometry of the planetary system? I submit that what Copernicus describes glowingly as a "wonderful symmetry in the universe, and a definite relation of harmony in the motion and magnitude of the orbs, of a kind not possible to obtain in any other way" is precisely that cosmographical

12. Regarding the uniqueness and supremacy of Trinitarian nondualism I refer to *Christian Gnosis* (Tacoma, WA: Angelico Press/Sophia Perennis, 2008), Chapter 7.

13. Isa. 65:17 and Rev. 12:1. Let me note that, mathematically speaking, Dante's integral cosmos constitutes a three-dimensional sphere, with the Empyrean as its (2-dimensional) equator. The Florentine poet appears to have been the first man to conceive of a more-than 2-dimensional sphere. On the cosmology of the *Divina Comedia* see also Titus Burckhardt, *Mirror of the Intellect*, op. cit., pp.17–26 and 82–98.

reflection. Admittedly, the Copernican and the Tychonian systems prove to be mathematically equivalent,[14] which is to say that they predict the same apparent orbits; even so the symmetries and harmony of which Copernicus speaks remain hidden in the Tychonian scheme, whereas they become resplendently manifest in the Copernican.

One has mixed feelings, therefore, concerning the contemporary defense of geocentrism. Whereas Christian believers are surely to be commended for guarding a doctrine basic to their faith, the reductionist spirit of the times has forced the debate onto a cosmographic plane on which the essential has already been lost, and where for that very reason the defenders find themselves at a distinct disadvantage. Admittedly the principle of relativity offers some protection to the beleaguered Tychonians, but at the cost of emasculating the geocentrist claim. Meanwhile the fact remains that a heliocentric coordinate system offers undeniable theoretical advantages precisely because it is adapted to the symmetries Copernicus had his eye upon: the very symmetries that bear witness to the heliocentric truth. The Tychonians may indeed be right in claiming that they too can explain the observable facts, but one wonders at what cost in the form of cumbersome *ad hoc* interventions.[15] One cannot but commiserate with these defenders whom the opposing side does not deem worthy even of a response.

What necessarily baffles the exoterist mentality is what might be termed the multivalency of authentic revelation, be it scriptural or cosmic. Truth is hierarchical, and so Holy Writ and the cosmos as such need be in a sense hierarchical as well. No single perspective or level of understanding—no single "*darshana*"—can do full justice to the integral truth: Revelation itself informs us of this fact in various ways. Typically both Scripture and the cosmic revelation do so by way of "fissures," that is to say, seeming incongruities which disturb

14. A sketch of the proof may be found in Thomas Kuhn, *The Copernican Revolution* (New York: MJF Books, 1985), pp. 201–206.

15. So far as planetary orbits are concerned, Tychonian astronomy is equivalent to the Copernican as we have said. *Ad hoc* interventions, however, are needed to account for stellar aberration and stellar parallax. See pp. 147–152.

and puzzle, and hopefully spur us on to seek a higher level of truth. As Christ Himself intimated to His disciples on the eve before His passion: "*I have yet many things to say unto you, but ye cannot bear them now.*"[16] Humility in the moral sense is not enough: we need also an intellectual and indeed theological humility. To preserve ourselves from falling into some arid dogmatism, we need ever to continue on our way: "from truth that is lower to truth that is higher." Dogmas, it seems, are meant for the *viator*, the spiritual traveler, not for the armchair theologian. It is not that dogmas of a sacred kind are simply provisional or limited in the ordinary sense, but rather that they harbor unsuspected truths. We need, as I have said, to continue on our way; as the author of Hebrews points out: "*Strong meat belongeth to them that are full of age.*"[17] And moreover, since truth derives ultimately from God, this step-by-step ascent constitutes indeed an *itinerarium mentis in Deum*, a veritable "journey into God." But clearly, it is an *itinerarium* in which the *viator* himself is progressively changed; in the words of St. Paul: "*But we all with open face beholding as in a glass the glory of the Lord, are changed into the same image from glory to glory, as by the Spirit of the Lord.*"[18]

Getting back to cosmography: the higher truth of heliocentrism, as I have said, is reflected in the superior beauty or "symmetry" of the corresponding mathematical description; but we need to remember that the "high truth" in question pertains to what may indeed be characterized as an esoteric level of vision. Reduced to a mere cosmography, heliocentrism ranks in reality *below* its geocentric rival; for the latter, insofar as it corresponds to the testimony of cognitive sense perception, opens upon vistas of truth, as we have noted before, which are inaccessible to the physical scientist as such. The problem with an "exoteric" geocentrism, on the other hand—a geocentrism that simply denies the heliocentric truth—is that in the final count it lacks a credible defense against a scientific heliocentrism: referents and epicycles, figuratively speaking, do not stand up well against the equations of Kepler and Newton. Even the most

16. John 16:12.
17. Heb. 5:14.
18. 2 Cor. 3:18.

committed geocentrist can hardly fail to recognize a superior cogency in the heliocentric theory, and secretly sense that some other truth must stand at issue, a truth not comprehended from the geocentric point of view. But alas, to the exoterist mentality that "other truth" is perforce hostile, an erroneous teaching that threatens the integrity of the geocentric worldview. What by right should spur us on to seek a higher, more comprehensive level of understanding—what *de jure* should be *liberating*—comes thus to be feared and rejected as a rank heresy.

What further complicates the issue is the fact that heliocentrism has generally come to be identified with the Galilean doctrine, which *is* in fact a rank heresy. I have already suggested that Galilean heliocentrism erodes the sense of verticality which supports and indeed enables the spiritual life: that it plunges us into a flattened and de-essentialized cosmos in which the claims of religion cease to be credible. I propose now to consider yet another ill effect of the Galilean heresy, which in a way is complementary to the aforesaid loss of verticality.

Every religion is perforce *homocentric* in its worldview. To put it in Christian terms: man occupies a central position in the universe because he is made in the image and likeness of Him who is in truth the absolute center of all that exists. Furthermore, man is central because, as the microcosm, he in a way contains within himself all that exists in the outer world, even as the center of a circle contains in a sense the full pencil of radii. Or again: man is central because he is the most precious among corporeal beings. Genesis teaches in fact that God created the Earth as a habitat for man, and the Sun, Moon, and stars *"for signs, and for seasons, and for days, and years."* It is on account of man's centrality, moreover, that the Fall of Adam could affect the entire universe. Now, it is true that the centrality of which we speak is above all metaphysical, or mystical as one might also say; yet even so, it is in the nature of things that this "essential" centrality should be reflected cosmographically. Does not the outer manifestation invariably mirror the inner or essential reality? To suppose that man can be metaphysically central while inhabiting a speck of matter occupying some nondescript position in some nondescript galaxy—that would surely be incongruous in the extreme. Once again: it would deny the very principle of cosmic symbolism, and thus the theophanic nature of cosmic reality. To be sure, it is possible to affirm metaphysical centrality on an abstract philosophic plane and in the same breath affirm cosmographic acentrality as well; I doubt, however, that one can do so on an existential level, that is to say, in point of actual credence. To the extent that we truly believe the stipulated acentrality of the Earth, we are bound to relinquish the traditional claim of homocentrism: in reality, I say, these two articles of belief are mutually exclusive. One can, I say, pay lip-service to both, as in fact a contemporary theologian may well do; but actual belief—that is something else entirely.

The objection may be raised that it is actually possible to espouse an acentric cosmology without detriment to the rightful claims of religion; and one might point to Nicholas of Cusa by way of substantiating that contention. True enough! One needs however to understand that the Cusan cosmology is profoundly Platonic, and corresponds indeed to an authentically esoteric point of view. Its so-called acentrality is consequently worlds removed from the Einsteinian, and could more accurately be described as a "pancentrality." The Cardinal, thus, does not simply deny the geocentrist claim, as does the Galilean astronomer: in reality he transcends the geocentrist contention, and in so doing, justifies and founds it "in spirit and in truth," paradoxical as this may seem. "It is no less true," declares Nicholas of Cusa, "that the center of the world is within the Earth than that it is outside the Earth"; for indeed, "the Blessed God is also the center of the Earth, of all spheres, of all things in the world."[19] Here, in this terse and lucid statement worthy of a sanctified mind, we breathe the pure and invigorating air of a Christian esoterism. It is ever the way of authentic esoterism to "deny" only by

---

19. *On Learned Ignorance*, trans. Jasper Hopkins (Minneapolis: Banning, 1985), p. 115. A masterful discussion of the Cusan cosmology may be found in Jean Borella, *The Secret of the Christian Way* (Albany, NY: State University of New York Press, 2001), chapter 2.

affirming a higher truth, which contains but yet vastly exceeds the original claim.

It is true that the Earth enshrines the center of the universe; but so do the Sun, the Moon, and the myriad stars. Yet it is evidently the first of these recognitions that matters most to us so long as we are denizens of this terrestrial world. As I have noted before, we depend upon that recognition, that truth, for our orientation: our spiritual orientation no less than our physical.

What happens, now, when we ascend from a geocentric to an authentically heliocentric worldview: do we retain the original homocentrism? One may surmise that as we transcend the geocentric outlook, we likewise transcend the lesser theological conception of homocentrism, in accordance with the Pauline dictum: "*I live, yet not I, but Christ liveth in me.*"[20] The resultant and indeed higher homocentrism is thus in reality a Christocentrism; but again, that Christocentrism is not destructive of the earlier notion, the lesser truth—even as the Christ who "*liveth in me*" is not destructive of the "I" that "*lives.*" It is once again a question of levels, of hierarchy. Meanwhile the intrinsic connection between geocentrism and the lesser homocentrism endures on the plane to which either notion applies, which is none other than that corresponding to our human condition. Let no one therefore deny either of these notions, these truths, "from below": the consequences of that denial cannot but be tragic in the extreme. Such a denial affects and indeed "poisons" every aspect of human culture, beginning with the life of religion, which it undermines.

It would be hard to overestimate the impact of the Copernican Revolution upon Western culture. Already in 1611, when that revolution had barely begun, John Donne appears to have divined its larger significance: "And new philosophy calls all in doubt," he laments; "'Tis all in pieces, all coherence gone." No wonder the ecclesiastical guardians of the Roman Church were apprehensive as well, without

20. Gal. 2:20.

perhaps realizing in full clarity what it is that ultimately stands at issue. Today, four centuries later, what lay concealed in that beginning has become clearly manifest, for all to see; as Arthur Koestler notes, it is "as if a new race had arisen on this planet." Could this be the reason why St. Malachi, in his famous prophesies, characterizes the reign of Pope Paul V (1605–1628) by alluding to the birth of "a perverse race"? One needs to recall that what is sometimes termed the first Galileo trial took place in the year 1616. What, then, could be that "perverse race" to which the saintly prophet refers? Given that Galileo is indeed "the father of modern science," one is compelled to answer that it is none other than the race of modern scientists, and by extension, the community of individuals imbued with the modern scientistic outlook. This, then, constitutes the fateful "birth" which took place during the pontificate of Paul V: no wonder St. Malachi singles out that event! It was not simply a question of planetary astronomy, obviously: what came to birth was indeed a "new philosophy" as John Donne was quick to realize. From that point onwards, Western man began to look upon the universe with different eyes; and thus he found himself, quite literally, in a new world. Goethe, as always the realist, surely did not overstate the case when he declared that "probably not a single fact has had a deeper influence on the human spirit than the teaching of Copernicus." Only one should add that whereas Copernicus proposed the heliocentric hypothesis—the new mathematical model, if you will—it was Galileo who supplied the new philosophy.

As everyone knows, Galileo was formally tried in 1633 and forced to recant his Copernican convictions. The proposition that the Sun constitutes the immobile center of the universe was declared to be "formally heretical, because it is expressly contrary to the Holy Scriptures." And so the matter stood until 1822, when, under the reign of Pius VII, the Church commenced to soften its stand with regard to what it termed "the general opinion of modern astronomers." Thus began a process of accommodation with "the new race" which came to a head in 1979, when Pope John Paul II charged the Pontifical Academy of Sciences to re-open the Galileo case, and if need be, to reverse the verdict of 1633. Given the mentality which came to the fore in the wake of Vatican II, the outcome of that inquiry was never in doubt: Galileo was exonerated—some would say, "canonized"—following which Pope John Paul II in effect apologized to the world for wrongs committed by the Church. Could this be the reason, perhaps, why St. Malachi alludes to this Pope in the enigmatic words *"De Labore Solis"*? To be sure, the phrase, which traditionally refers to the movement of the Sun, does relate to Galileo, the man who denied that the Sun does move. Could it be, then, that St. Malachi, having previously signaled the birth of a "perverse race," is now alluding to the fact that some four hundred years later the Church has reversed its stand and relinquished its opposition to that "race," which is to say, to that new philosophy? Certainly St. Malachi's allusion can be interpreted in other ways as well; for example, *"De Labore Solis"* might be taken as a reference to the fact that this Pope, who has traveled far more extensively than any of his predecessors, has so many times "circled the globe" in his papal airliner (named "Galileo," interestingly enough).

But be that as it may, the fact remains that the Church has now joined the rest of Western society in adopting a scientistic worldview; during the reign of Pope John Paul II, and obviously with his sanction, a Copernican Revolution has finally taken place within the Church itself. Yet, to be precise, it is not the Church as such that has undergone change—that has "evolved" as the expression goes—what has changed, rather, is the orientation of its human representatives: it is Rome, let us say, that has reversed its position. Humanly speaking, the ecclesiastic establishment may have opted for the only viable course: given the sophistication and prowess of contemporary science—given the *"great signs and wonders"* that could *"deceive even the elect"*—it may not be actually feasible to stem the mounting tide of scientistic belief. One must nonetheless insist, in light of the preceding analysis, that the contemporary rejection of geocentrism is not in fact compatible with Christian doctrine. To the extent, therefore, that Rome has embraced that position, it has compromised the true teaching of the Church: this is the crux of the matter.

9

Intelligent Design
and Vertical Causality

Now, whatever lacks intelligence cannot move towards an end,
unless it be directed by some being endowed with knowledge
and intelligence, as the arrow is shot to its mark by the archer.
St. Thomas Aquinas

From time immemorial mankind has understood events or objects as the result of necessity, or of chance, or of design. These three basic categories of explanation appear to be native to the human mind, and as such they constitute what may be termed pre-philosophical notions. To be sure, there has been an ongoing effort on the part of philosophers to clarify these conceptions, and integrate them into a coherent account of causality. The simplest approach, perhaps, is to deny both chance and design as the Greek atomists have done, and thus to suppose that all things occur by force of necessity as Leucippus declared. Other schools, while still denying chance, have acknowledged design as a principle of causality not reducible to necessity; such was the case in the Stoic philosophy, which stipulated a kind of providential action or *pronoia* emanating from the World-Reason known as the Logos. Yet other schools of thought acknowledge both chance and necessity as irreducible principles, but deny design; and it is of interest to note that most scientists of our day seem to espouse that position.

Among the three basic categories of causation, the most puzzling is perhaps the notion of chance. What tends to confuse the issue, first of all, is the mistaken yet commonly held belief that chance and necessity are mutually exclusive: that it is a question simply of "either or." But clearly, to say that the toss of a coin yields heads or tails "by chance" is not to claim that the outcome has no cause, or is not in fact determined by its cause. Whether or not the toss of a coin is deterministic in some ultimate sense is a separate issue; what counts is that the event is, in any case, random or contingent in a suitably relative sense. Thus one finds that even in the heyday of classical physics, when the operations of Nature were deemed to be fully deterministic, statistical methods based upon the idea of chance could be successfully applied in various domains: for instance, in the kinetic theory of gases. The random distribution of gas molecules and their velocities within a statistical ensemble, thus, does not contradict the notion that the trajectory of each molecule is fully determined by a causal law. And I would point out that this accords with Aristotle's idea that chance refers to the coincidence of causally determined sequences of events, a scenario which occurs, for example, when two classical particles collide. To be sure, the notion of chance has played an essential role in civil life as well: courts of law, for example, distinguish regularly between accidental and non-accidental happenings, and insurance companies treat catastrophic events as a random variable with a probability distribution which it is needful for them to know. It is definitely meaningful, therefore, to speak of contingency and likelihood whether Nature proves ultimately to be deterministic or not, or whether the outcome of every process is perhaps known beforehand to God. After all, we judge of things on the basis of *our* knowledge, and from our point of view; and as that knowledge, or that point of view, changes, so do our judgments relating to causality.

As concerns traditional doctrine, I contend that no orthodox school has ever been averse to the notion of chance. I have in fact argued on traditional ground that there *can be* no such thing as a fully deterministic universe, that contingency constitutes the complementary aspect of determination, the *yin*-side of the coin.[1] It is needful, then, that there be contingency as well as law. The inherently Cartesian concept of a clockwork universe turns out to

---

1. *The Quantum Enigma,* op. cit., pp. 100–103.

be fatally flawed; and surprisingly enough, this discovery has been made in our day, not by poets or mystics, but by mathematical physicists no less. One might think that these scientists, fully dedicated as they are to the ideals of rigor and exactitude, would be the last to reach that conclusion! If then such as these have declared contingency to be necessary after all, this finding carries weight.

In light of these considerations one may combine necessity and chance into a single category under the title of natural causation, which needs now to be distinguished from *design*. And again we find that these two modes of causation do not stand in opposition: that it is not a question of "either or." It proves in fact to be an essential characteristic of traditional cosmology to admit both modes: the divine, if you will, as well as the natural. It would thus be as contrary to the wisdom of tradition to maintain that events and objects in the natural world are caused exclusively by divine action—as certain religious extremists have claimed—as it would be to imagine that the universe is governed simply by natural causes. In sharp contrast to the philosophy of naturalism, traditional cosmology acknowledges two seemingly opposed principles: the primacy of divine action, namely, and the efficacy of natural causes.

I should perhaps remark that whereas Etienne Gilson may well be right in claiming that the harmonization of these two principles has found its consummation in the Thomistic philosophy, I cannot accept his contention that Platonism denies the efficacy of natural causes and thus affirms a radical extrinsicism.[2] To be sure, the philosophies of Plato and Aquinas represent different points of view, and it may be true that the efficacy of natural causes is affirmed more prominently in the writings of the Angelic Doctor; yet even so I find Gilson's charge of "radical extrinsicism" to be misplaced. The moral, perhaps, is that no one, however brilliant, can understand a traditional philosophy who does not approach that subject with reverence, or as I like to say, "with folded hands." To reiterate: *all* traditional cosmology respects, in the final count, both the primacy of divine action *and* the efficacy of natural causes.

2. *The Christian Philosophy of St. Thomas Aquinas* (Notre Dame, IN: University of Notre Dame Press, 1994), p. 185.

It has been further recognized that manifest design cannot be attributed to natural causation. This is in fact what St. Thomas Aquinas contends in his fifth proof for the existence of God; and let us note that the argument is validated, not by some abstract logic, but indeed on metaphysical ground, and thus on the basis of an intellectual perception. With the advent of modern times, however, that "argument from design" has come under attack. First came deism, the "absentee landlord" philosophy which in effect exiled God from the universe; and this has led, by stages and degrees, to the full-fledged naturalism that came into vogue during the nineteenth century. As one dictionary of philosophy puts it: "Naturalism holds that the universe requires no supernatural cause and government, but is self-existent, self-operating, and self-directing; that the world-process is not teleological and anthropocentric, but purposeless and deterministic, except for possible tychistic events." Among the various philosophical and theological movements which opposed this position, it was the school of British natural theology that centered its counterattack on the argument from design. "During the 17th and 18th centuries," we are told by Vergilius Ferm,

> there were attempts to set up a 'natural religion' to which men might easily give their assent and to offset the extravagant claims of the supernaturalists and their harsh charges against their doubters. The classical attempt to make out a case for the sweet reasonableness of divine purpose at work in the world was given by Paley in his *Natural Theology*, published in 1802.

Despite the fact, however, that this "natural religion" may have held its attraction for many an English gentleman, one finds that it fell woefully short of a tenable doctrine due to the fact that it had in part assimilated the very naturalism it wished to combat. In a word, British natural theology was a compromise solution, an eclectic doctrine that was bound to fall. And no one, it seems, knew better how to expose and capitalize on its weaknesses than Darwin himself: "I cannot persuade myself," he wrote, "that a beneficent and omnipotent God would have designedly created the Ichneumonidae [parasitic wasp] with the express intention of their feeding within the living bodies of Caterpillars."[3] True enough: there can be no answer to Darwin's objection, nor to the allied argument from dysfunction, without committing in some way to the doctrine of Original Sin and the resultant Fall—something British natural theology, in its "sweet reasonableness," neglected to do. And so it came about, in the wake of Darwin's theory, that this "natural theology," which had enjoyed the approbation of an intellectual and social elite, succumbed eventually to the assault of naturalism.

We must not however lose sight of the fact that there is substance and indeed validity in Paley's conception of a "watchmaker God." If we walk through a field, as Paley invites us to do, and discover a watch lying on the ground, we may indeed conclude that this object is not the product of natural causes: it was not a blind concatenation of accidental happenings that fashioned and assembled the parts of the watch to the end of keeping time. We are all perhaps familiar with a book entitled *The Blind Watchmaker,* in which Oxford zoologist Richard Dawkins proposes to refute Paley's claim; meanwhile however it turns out that the ancient argument has been recently revived on a scientific plane. The movement was sparked by Michael Behe, a molecular biologist, who introduced the concept of what he termed "irreducible complexity" and argued that no natural process can give rise to structures that are in fact irreducibly complex. In *Darwin's Black Box,* published in 1996, Behe took his case before the general public. His treatise, filled with captivating accounts of discoveries from the world of molecular biology, has become widely known, and has engendered serious debate in scientific circles. Behe's book, however, was only the beginning: the opening salvo, one might say, of a *scientific* counterattack, this time, against the prevailing naturalism. The decisive breakthrough was achieved two years later by a mathematician and philosopher named William Dembski, in a treatise entitled *The Design Inference.* Dembski had asked himself whether design can perhaps be recognized by means of some signature, some criterion which can be defined in strictly mathematical terms, and the resultant theory not

3. Francis Darwin, ed., *The Life and Letters of Charles Darwin* (London: Murray, 1887), pp. 303–312.

only generalizes Behe's concept of irreducible complexity, but puts the question of a "design inference" on a mathematical—and hence rigorous—basis. What Dembski discovered is that a signature or criterion of design can indeed be given in terms of a probabilistic notion of *specified complexity,* or equivalently, in terms of an information-theoretic concept of *complex specified information* or CSI. The decisive result is a conservation theorem for CSI, which affirms in effect that CSI cannot be generated by any natural process, be it deterministic, random, or some combination of the two, as in so-called evolutionary algorithms. Thus a new science termed "design theory," also known as the theory of "intelligent design" or ID, came into existence. The movement has of course drawn sharp criticism from various segments of the scientific establishment, beginning with the Darwinist contingent; yet it is hard to argue with a mathematical theorem, at least so long as one plays by the rules.

Let us now take a closer look at ID theory, beginning with the concept of irreducible complexity: "By *irreducibly complex*," writes Behe, "I mean a single system composed of several well-matched parts that contribute to the basic function, wherein the removal of any one of the parts causes the system to effectively cease function."[4] The definition is evidently framed "with malice of forethought" to guarantee that no Darwinist process can ever give rise to an irreducibly complex structure; for as Darwin himself observed: "If it could be demonstrated that any complex organism existed, which could not possibly have been formed by numerous, successive, slight modifications, my theory would absolutely fail." To which however he added: "But I can find no such case."[5] The logic of Behe's argument appears to be impeccable: if a structure requires a number of "well-matched parts" before it can be functional, this precludes the possibility that "numerous, successive, slight modifications" could have been successively selected on the basis of function. The viabil-

4. *Darwin's Black Box* (New York: The Free Press, 1996), p.39.
5. *On the Origin of Species* (Harvard University Press, 1964), p.189.

ity of Darwin's theory, therefore, does indeed hinge on the question whether one can "find such a case." Now, one of the most impressive and frequently cited examples of irreducible complexity is the so-called bacterial flagellum: a molecular device, whose function it is to propel the bacterium through its watery environment upwards along a nutritional gradient. The device consists of an acid-powered rotary engine—replete with a rotor, a stator, O-rings, bushings, and drive shaft—plus the actual flagellum, a kind of molecular rotary paddle. On account of the disorienting effect of Brownian motion, the flagellum must rotate at angular velocities on the order of 10,000 rpm, and must be able to reverse direction within one hundredth of a second. Moreover, to be functional the device obviously requires auxiliary structures for detection and control, as well as for the production, storage, and distribution of the requisite fuel. What confronts us here, quite clearly, is a feat of nanotechnology that staggers the imagination; and needless to say, no one has yet proposed so much as the vaguest outline of a Darwinist scenario that might account for the production of these structures.

Yet, even so, Behe's argument remains incomplete. The conclusion that no Darwinist process could have produced the bacterial flagellum does appear to be inescapable; and yet the argument falls short of a rigorous proof. Now, the standard strategy, in the physical sciences, for proving that a closed system, operating under the action of natural forces, cannot attain a certain state involves the use of an invariant satisfying a conservation law. It matters not whether the invariant is an energy, for example, which must remain constant, or a quantity that can change only in one direction, such as entropy: in either case the law in question evidently rules out states which would violate that law. And let us note that this argument does not require that we check out all possible scenarios that might conceivably bring the system into the disputed state: the law itself settles the matter at one stroke. Getting back to Behe's contention: it is by means of this strategy, using CSI as the invariant and Dembski's theorem as the requisite law of conservation, that one is able—for the first time!—to refute the Darwinist claim, in the case of structures such as the bacterial flagellum, *with mathematical precision.*

The basic idea of Dembski's theory is simple enough. Let us suppose that someone shoots arrows at a wall. To conclude that a given strike cannot be attributed to chance—in other words, to effect a design inference—one evidently needs to prescribe a target or bullseye that sufficiently reduces the likelihood of an accidental hit. What is essential is that the target can be specified without reference to the actual shot; it would not do, for example, to shoot the arrow first, and then paint a bull's-eye centered upon the point where the arrow struck. What stands at issue, however, has nothing to do with a temporal sequence of events: it does not in fact matter whether the target is given before or after the arrow is shot. What counts, as I have said, is that the target can be specified without reference to the shot in question. In Dembski's terminology, the target must be "detachable" in an appropriate sense.

Consider a scenario in which the keys of a typewriter are struck in succession. If the resultant sequence of characters spells out, let us say, a series of grammatical and coherent English sentences, we conclude that this event cannot be ascribed to chance. An exceedingly "small" and indeed "detachable" target has been struck, which however was, in this case, specified *after* the event. In general, the specification of a target requires both knowledge and intelligence; one might mention the example of cryptanalysis, in which specification is achieved through the discovery of a code. What at first appeared to be a random sequence of characters proves thus to be the result of intelligent agency. The fact is that it takes intelligence to detect intelligent design.

I would like to emphasize that it is impossible to rule out the hypothesis of chance simply on the basis of low probability. If a sequence of 1's and 0's is generated by tossing a fair coin a billion times, the possibility that the resultant bit string will contain not a single 0, say, can indeed be validly ruled out. Yet, if one does actually toss a coin a billion times, one produces a bit string having exactly the same probability as the first: one half to the power one billion, to be precise. Why, then, can the first sequence (the one containing no 0's) be ruled out, whereas the second can not? The reason is that the first conforms to a pattern or rule which can be defined independently; it is a question, once again, of a "detachable" target which itself has low probability. In the case of the first sequence, the prescription "no 0's" in itself defines a target of that kind: the subset, namely, containing the given bit string and no other. But this is precisely what can *not* be done in the case of the second bit string (the one produced by tossing a coin a billion times): it is virtually certain, in that case, that no *detachable* target of low probability has been hit. It is possible, of course, to produce a target from the empirical sequence itself; but that description or pattern (if such it may be called) would turn out *not* to be detachable. It would be comparable to a bull's-eye painted around the spot on the wall where an arrow has struck: such a description, of course, proves nothing. The discovery of a *detachable* pattern of sufficiently *low probability*, on the other hand, proves a great deal: it may prove, in fact, that the event in question cannot be attributed to chance. Thus, what rules out chance is not low probability alone, but low probability in conjunction with a detachable target: this winning combination is what Dembski terms *the complex specification criterion*.

How then does one formalize these considerations to arrive at a rigorous mathematical theorem that encapsulates the decisive recognition? I will indicate how this is done, with the understanding that the "non-mathematical" reader may skip to the last paragraph of this section, which summarizes the implications of Dembski's discovery.

The mathematical structure within which design theory operates is as follows: One postulates an "event space" which is simply a set Ω, together with a probability measure P that assigns to each measurable subset of Ω a real number between 0 and 1. A measurable subset of Ω is now an *event*, and a point an *elementary* event. Given an elementary event E, a *specification* of E is a subset T of Ω containing E, which is *detachable* in a technical sense, the definition of which I will omit in this summary. The pair (T, E) is then said to constitute a *specified event*. It is to be noted that a specified event consists of two components: a *conceptual* component T, one can say, and a *physical* component E. It constitutes thus a twofold entity, a thing that combines, so to speak, two worlds. And therein, let me add, lies the power and indeed the genius of Dembski's theory: where others have dealt simply with events, Dembski has his eye on *specified* events: a categorically different kind of entity.

Let us suppose, now, that one is able to associate an invariant I with each specified event which satisfies a conservation law: which cannot, say, increase under the operation of natural causes. As we have previously noted, such a law can rigorously validate a theory of design. Now, to obtain such an invariant Dembski replaces the probability measure P by a corresponding *information* measure I, defined by the equation

$$I = -\log_2 P.$$

Let us note that the information contained in a bit string of length n is just n, and that for any event A in Ω, $I(A)$ represents the information content of A *as measured in bits*. It is to be noted that P and I are inverse measures: the smaller P, the larger I will be; and indeed, as P tends to zero, I tends to infinity. In mathematical parlance, I is thus a measure of *complexity*. Dembski next defines the information content of a *specified* event (T,E) to be $I(T)$: what counts, in other words, is the *conceptual* component of the specified event. And finally he defines *specified information* as the information contained in a specified event: and this is what he takes initially as his invariant.

It turns out, first of all, that *specified information is strictly conserved under the action of a deterministic process*. And as one might expect, the proof hinges upon the fact that a deterministic process can be represented by a function f having the event space Ω_0 of initial states as its domain and the event space Ω of the resultant states at a later time t for its range. Given an elementary event E_0 in Ω_0, let $E = f(E_0)$. So too, given a specified event (T,E) in Ω, one can define T_0 as $f^{-1}(T)$ to obtain a corresponding specified event (T_0, E_0) in Ω_0. Similarly, given a probability measure P_0 on Ω_0, one obtains a probability measure P on Ω by taking $P = P_0 f^{-1}$. It follows that the specified events (T_0, E_0) and (T,E) carry precisely the same amount of specified information, as was to be shown.

Surprisingly, this conclusion could have been foreseen without recourse to mathematical analysis; as Dembski has put it: "What laws cannot do is produce contingency; and without contingency they cannot generate information. . . ."[6] The point was in fact made four decades earlier by Leon Brillouin when he wrote that "a machine does not create any new information, but it performs a very valuable transformation of known information"; and what is perhaps most amazing of all, as far back as 1836 the poet and amateur scientist Edgar Allan Poe had said much the same.[7]

But what about *random* processes? As we have seen, a random process can generate arbitrarily large amounts of information (due to the fact that events of arbitrarily small probability can indeed occur), and can even generate *small* amounts of specified information. What it cannot do, according to the complex specification criterion, is generate specified information in *large* amounts: that is the crucial point. The proverbial monkey pounding on a typewriter can perhaps produce a few bits worth of English prose, but definitely not the text of Hamlet. There must consequently be a cut-off, even though its exact location cannot be specified. What *can* be specified—though not uniquely!—is a so-called universal complexity bound or UCB beyond which specified information cannot be increased by a random process within the range of physical space and time. Dembski for good reason takes his UCB to be 500 bits, which is "playing it safe": no monkey could remotely do that well, not in a billion years! What no random process can generate, therefore, turns out to be *complex* specified information or CSI: specified information, namely, in excess of the UCB.

Having thus disposed of deterministic as well as random processes, it remains to consider an arbitrary combination of the two, which is what one terms a stochastic process.[8] In place of a func-

---

6. *No Free Lunch* (New York: Rowman & Littlefield, 2002), p.155. It should perhaps be mentioned that the expression "no free lunch" has of late acquired a technical sense: it has come to refer to a class of theorems regarding so-called "evolutionary algorithms," proved in the latter part of the 90's, which are expressive of the fact that the problem-solving capacity of such algorithms proves to be severely limited. Dembski's theory can be seen as a generalization of these "no free lunch" theorems.

7. See Leon Brillouin, *Science and Information Theory*, 2nd ed. (New York: Academic Press, 1962), pp. 267–69.

tion f(x) of a single variable, one has now a function $f(x, \omega)$ of two variables, in which ω represents the random component of the process. The trick is to break the problem into two parts: one first permits ω to "occur," which transforms the original function f into a function f_ω of a single variable x, in which ω serves as a fixed parameter. That function, however, can then be handled as in the deterministic case, which leads finally to the conclusion that the total process cannot increase specified information by more than the UCB.[9]

So much for our—duly simplified—summary of Dembski's proof. What concerns us now is the significance of this marvelous mathematical theorem, this universal conservation law for CSI: what exactly does it tell us in regard to evolutionary biology? The theorem asserts that CSI cannot be generated by any natural process, which manifestly implies that the vast amounts of CSI existing within the DNA of every living cell could not have been produced through the Darwinist scenario of random variations acted upon by natural selection. But could there possibly be a way out of this dilemma? Neo-Darwinists such as Manfred Eigen, having recognized the origin of information as the major problem confronting contemporary biology, have been searching for an evolutionary algorithm that could accomplish the needful, without however realizing that this possibility can in fact be ruled out on theoretical grounds: for it happens that every evolutionary algorithm, no matter how ingeniously conceived, reduces perforce to a stochastic process, and consequently falls under Dembski's interdict. Only two possibilities remain: either the universe must have been replete with CSI from the first moment of its existence—a supposition which hardly

8. The classic example of a stochastic process is the so-called Brownian motion of small particles immersed in a fluid, which gives rise to a Newtonian trajectory modified by discrete random collisions with high-speed molecules. The physics of Brownian motion was successfully given by Albert Einstein himself in a paper published in 1905, the same year in which he explained the photoelectric effect and brought out what is known as the special theory of relativity.

9. The full proof may be found in William A. Dembski, *No Free Lunch*, op. cit., pp. 149–166.

squares with the big-bang hypothesis—or else it cannot be conceived as a closed system operating under the action of natural causes.

It emerges that CSI, and more generally, *information* in the mathematical sense, is something irreducible, that it has an existence of its own. It was Norbert Wiener, I surmise, who first pointed out that in addition to mass and energy, the universe comprises "information" as a basic ingredient. It is in any case to be noted that from about 1900 onwards the concept of information has come to occupy an ever more prominent place in the physical and biological sciences. The trend began with Boltzmann's statistical mechanics in which the twin notions of contingency and probability play a vital role. Contingency and probability, however, add up to information as we have seen; and let us note that the statistical definition of entropy as given by Boltzmann is actually formulated in information-theoretic terms. It was Boltzmann's statistical approach to blackbody radiation, moreover, that enabled Max Planck to discover the so-called quantum of action, nowadays known as Planck's constant, a discovery which inaugurated the quantum era.[10] And with the advent of quantum mechanics it became apparent that on the most basic level of physical theory probabilities and information are not only useful conceptions but prove to be *necessary* as well.

It is true that Einstein, for one, refused to accept this conclusion; yet every argument which he advanced to counter the quantum-mechanical indeterminism proved ultimately to be ineffectual. More often than not, it actually contributed insights which served in the end to bolster the quantum-mechanical position, as in the case of the well-known Einstein-Podolsky-Rosen thought experi-

10. Having arrived at the correct formula for blackbody radiation by empirical means, Planck needed Boltzmann's statistical theory to arrive at its physical significance. As he put it in his Nobel Prize address of 1920: "After some weeks of the most intense work of my life, light began to appear to me and unexpected vistas revealed themselves in the distance."

ment, which was in fact eventually carried out, with the result that Einstein, *et al.*, were mistaken. Moreover, when David Bohm, after prolonged discussions with Einstein, did finally succeed in constructing what appeared formally to be a deterministic quantum theory, he did so at the cost of introducing what he termed *active information* as a basic principle. It turns out, moreover, as Dembski was quick to point out, that Bohm's "active information" is but a special case of CSI! One sees in retrospect that Bohm was able to dispense with indeterminacy on the level of particles only at the cost of admitting contingency in the form of CSI: one way or another, it seems, contingency—and thus information—is bound to enter the picture. Meanwhile CSI has made a second appearance as a foundational concept of physics, this time in the form of the so-called Fisher information from which Roy Frieden was able to derive all the basic laws.[11] From yet another direction entirely, recent advances in communication and computer technology have given rise to a number of mathematical sciences in which information, normally in the form of CSI, plays a central role. Information theory itself, as a matter of fact, was inaugurated as a mathematical discipline in 1948 by Claude Shannon, an electrical engineer concerned with communication problems. But unquestionably the most significant encounter with information has taken place in the domain of molecular biology, which has uncovered what may be termed *the primacy of CSI in the biosphere.* The fact that vast quantities of specified information, recorded in a four-letter alphabet, reside within every living cell, and that each species derives, as it were, from a text known as its genetic code—this fact does indeed leave little doubt in that regard.

Given the crucial role of CSI in both physics and biology, it behooves us now to reflect further upon that notion, beginning with the mathematical concept of information as such. The danger,

---

11. *Physics from Fisher Information* (Cambridge University Press, 1998). I have dealt with this question at some length in Chapter 2.

when it comes to the latter, is that we are prone to read more into the technical term than it is meant to signify: the word had after all been in use for a very long time before Shannon bestowed upon it a technical sense. That sense is in fact rather bare: it boils down to the actualization of an event represented by a subset E in a mathematical space with probability measure P. If I flip a coin n times I have produced information: n bits worth, to be exact. And even now, as I am striking the keys of my keyboard, I am producing Shannon information. I am also, however, generating *semantic* information, which is something else entirely: something, in fact, which no mathematical theory can ever encompass for the obvious reason that semantic information does not reduce to quantity, to mere sets and relations. There is an ontological discrepancy between semantic and Shannon information, not unlike the ontologic hiatus separating the corporeal and the physical domains. And just as a corporeal object X determines an associated physical object SX, so also does every item of semantic information determine a corresponding item of Shannon information which serves, so to speak, as its material base: the latter is simply what remains when all that is non-quantitative has been cast out or "bracketed." One thus arrives, once again, at René Guénon's crucial point that "quantity itself, to which they [the moderns] strive to reduce everything, when considered from their own special point of view, is no more than the 'residue' of an existence emptied of everything that constitutes its essence."[12]

Having distinguished between semantic and Shannon information, it should be noted that the semantic component constitutes a *specification* in Dembski's sense, and in fact defines a *detachable* target.[13] To be sure, the example of semantic information is highly special, which is to say that specification can arise in a thousand other ways. Think of a bit string in which 1's and 0's alternate, or in which they represent a sequence of prime numbers in binary notation; or

---

12. *The Reign of Quantity* (Hillsdale, NY: Sophia Perennis, 2004), p. 5.

13. As the reader may have observed, we have, in our summary, left the definition of "detachability" out of account. Suffice it to say that it hinges upon the notion of "rejection functions," a technical concept with which we need not concern ourselves.

again, think of a bit string of length n which is "algorithmically compressible" in the sense that it can be generated by a computer algorithm of "length" less than n (a notion which can indeed be defined): all these are examples of specification. It appears, however, that despite its highly special nature semantic specification enjoys a certain primacy in the natural domain: if indeed God "spoke" the world into being as Scripture declares, such CSI or "design" as it carries must derive ultimately from a divine Idea or *logos*, which may by analogy be termed a *word*. And I would add that nowhere in the natural world is the linguistic character of specification more clearly in evidence than in the genetic code of an organism, which as I have noted before, constitutes a text recorded in a four-letter alphabet. The genetic code is thus a *written* text imprinted on DNA. Yet one may conjecture on theological ground that this written text derives from a *spoken* word: the kind to which Christ alludes when He testifies that *"the words I speak to you are spirit and life."*[14] No other scientific finding, I believe, is as profoundly reflective of theological truth as is the discovery of what may be termed the informational basis of life.

Following these somewhat cursory considerations relating to the concept of CSI, I now propose to reflect in some depth upon the nature of causality. We have so far distinguished between *necessity*, *chance*, and *design*, and have combined the first two modes under the heading of *natural* causation. On the strength of Dembski's theorem, moreover, we may conclude that CSI must be an effect of design. I now contend that this alternative mode of causation, which does not reduce to the natural, is none other than what I have elsewhere designated by the adjective "vertical," a mode characterized by its *atemporality*.[15] Inasmuch as vertical causation acts "above time" or instantaneously as one can also say, it differs fundamentally from the natural mode, which is inherently temporal, and

14. John 6:63.
15. *The Quantum Enigma*, op. cit., Chapter 6.

could therefore be characterized in metaphysical parlance as "horizontal." One arrives thus at a dichotomy which needs now to be carefully examined and clarified.

The prime example of vertical causation is unquestionably afforded by the creative Act of God; for as St. Augustine observes: "Beyond all doubt, God created the world, not in time, but with time." One needs however to realize that this creative Act extends in a sense to God's providence, the first effect of which is what theologians term "conservation." As Gilson points out, this effect "is, in some way, but the continuance of the creative act." We need not attempt to classify God's action upon the world; suffice it to say that God acts ever above time, and thus "vertically." That vertical causation, moreover, is the cause, not only of time, but also of the actions and processes that transpire in time. Yet these actions and processes have an efficacy of their own: such is the miracle of God's creation. God is intimately present, not only in the substance of all beings, but in their operations as well; and yet, as Gilson has beautifully put it: "The intimacy of the assistance He gives leaves their efficacy intact." It follows that God's vertical causality is complemented by a secondary causality which operates in time. And this is what we have termed *horizontal*: it is the kind, obviously, with which science is concerned. A word of caution, however, needs to be interposed: although every act of natural causality is horizontal—because it is effected by a temporal process—this does not imply that every act of horizontal causality is necessarily natural. There may conceivably be temporal processes which are neither deterministic, nor random, nor yet stochastic, and hence are not *natural* according to our definition, a fact which implies that the concept of horizontal causation may be wider than that of natural causality. We shall have occasion to return to this point presently.

Having distinguished between vertical causation, which is proper to God, and a horizontal causality which operates by way of a temporal sequence, we need to ask ourselves whether this dichotomy amounts simply to the traditional distinction between primary and secondary causation. The answer, clearly, is that it does not; for it happens that there exist second or created causes which likewise act above time, and thus vertically. What stands at issue is a higher degree of participation in the divine causality, one which preeminently reflects the action of God Himself. There are two prime examples of this higher mode of secondary causation: the causality, namely, emanating from the angelic realm, and in second place, action derived from human intelligence. Could it be that vertical causation is in fact the hallmark of intelligence? This appears indeed to be the case: to act "above time" is apparently the prerogative of an intellectual nature, a being endowed with intellect and free will. In short, whereas natural causation is "blind," it belongs to the very essence of *vertical* causation to be "intelligent."

This brings us to an issue which has been lurking in the background from the start: the question, namely, of human art in the widest sense of that term. To the contemporary mind at least, the most obvious and incontrovertible instances of "design" derive, not from an act of God, but from the action of a human artisan: not even the most committed Darwinist would deny the presence of design in the case of Paley's watch. But whereas he accepts the notion of "design" when it comes to human artifacts, he considers it "naïvely anthropomorphic" to extend this notion to the biological domain. We are in fact surrounded on all sides by CSI deriving from intelligent human action, and long before Dembski appeared on the scene, it was clear to everyone that what is now termed CSI was put there by an intelligent agent. Who, for example, when he comes upon a collection of stones on a hillside spelling out the word "Welcome" would imagine that these stones were left by a flood or an avalanche? In a thousand ways, all of us have been engaged, since early childhood, in the business of inferring design; and whether we know it or not, these inferences are invariably based upon *specification*. In fact, Dembski's theory merely formalizes—and in so doing, vastly generalizes—a thought process, indigenous to mankind, which enables us to recognize man-made "design."

Let us then consider the production of artifacts, from primitive crafts to modern industry. Is it not obvious, one might point out first of all, that an artifact is necessarily produced by means of a temporal process? In a sense this is of course true; I do not deny the necessity of temporal process: what I deny is its sufficiency. My contention is twofold: first, that the critical factor—the *sine qua non* of human art—is an act of intelligence; and secondly, that such an act is *not* reducible to a temporal process. Few, I suppose, would object to the first claim; it is the second that troubles us. The difficulty stems from the fact that we tend to temporalize the act of intelligence by identifying *cognition* with *thought*. We take it for granted that cognition occurs *within* thought—within a psychosomatic and temporal process—whereas in fact thought is only a means, a movement, if you will, in quest of cognition. To put it in traditional terms: cognition is an *intellective* as opposed to a *psychosomatic* act. As St. Thomas Aquinas observes: "The activity of the body has nothing in common with the activity of the intellect."[16] And this in itself suggests strongly that the intellectual act does not take place "in time." The crucial point, however, is that it *cannot* take place in time, for the simple reason that temporal dispersion is incompatible with cognition: we cannot know "bit by bit," because to know is necessarily to know *one* thing. This conclusion cannot be obviated, moreover, by adducing memory as a means of presentifying the past; the fact remains that the cognitive act must be "instantaneous," and hence supra-temporal: for indeed, the moment or instant is not a part of time. The intellect, therefore, whether conceived as a "third principle" in accordance with the tripartite *corpus/ anima/spiritus* anthropology, or (Thomistically) as a power of the soul, must in either case be inherently supra-temporal as well.

Getting back to the production of artifacts, the following has now become clear. If intellectual agency is indeed a *sine qua non* of human making, it is *ipso facto* impossible to reduce the production of the artifact to a temporal process. It is true, of course, that mechanized manufacture *is* a temporal process; but one must not forget that the machinery involved in this process carries design, which is transmitted to the resultant product. As Dembski's analysis shows, a deterministic process, and thus a function, may indeed *transmit* CSI, but cannot *produce* it. A machine-made artifact, thus, no less than one produced by a human artisan, presupposes an act of vertical causation.

16. *Opusculum, De unitate intellectus contra Averroistas,* iii, quoted in J. Rickaby, *Of God and His Creatures* (Westminster, MD: Carrol Press, 1950), p. 127 n.

Human making, amazingly enough, is allied to God's creative act by virtue of the fact that it likewise entails an *atemporal* mode of causation. In a way the human artist imitates the divine, and "participates" to some degree in God's creative agency: "Art imitates Nature [in the sense of *natura naturans*] in her manner of operation," says Aquinas.[17] In imitation of the Holy Trinity, no less, the human artist works "through a word conceived in his intellect."[18] It is evidently no small thing that transpires in even the humblest act of human making, which is indeed worlds removed from a mere temporal process: no wonder this categorical difference can be detected in the artifact by way of a distinctive signature indicative of design!

A further clarification needs now to be made. Having characterized vertical causation by the fact that it is atemporal in its mode of operation, we must bear in mind that it may nonetheless be temporal in its effect. A violinist, for example, does indeed act *qua* artist "above time" on the plane of intellect, and yet the music he plays is produced by a movement of his bow. Once again a temporal and thus horizontal process enters the picture; but clearly, it is no longer a *natural* process: it cannot be, since it springs from an *intellectual* act. One needs therefore to distinguish (as I have intimated previously) between two kinds of temporal process or horizontal causation: the kind that derives from natural causes alone, and the kind that springs from intelligent agency. It is the violinist, acting as an intelligent agent, who first apprehends the music—Dembski's "detachable pattern"—on the plane of intellect, and then, by an act of his free will, conveys that pattern to the world of sense by way of a temporal process, an action of horizontal causality.[19]

A few words on the subject of "free will" are called for at this juncture. One sees from the example of the human artisan that intelligent action *is not*—and cannot be—the result of a natural process. The cause of such productive action cannot therefore be situated on the natural plane, that is to say, in the external world,

17. *Summa Theologiae* I, Quest. 117, Art. 1.
18. Ibid., Quest. 45, Art. 6.

and may consequently be characterized—by default, if you will—as "internal." But is this not indeed what we normally mean when we speak of "free will"? Now, if by "freedom of the will" we understand an exemption from natural or external causality, then the preceding reflections do in fact establish that freedom in the context of human art. The decisive fact, however, is that intelligent action is "free" by virtue of an intimate participation in the freedom of God Himself. And to be sure, this divine Freedom is infinitely more than a mere exemption from external constraint: after all, it is by virtue of this very Freedom that God created the world, and thus "external constraint" itself. Freedom, therefore, is not simply exemption from external constraint: it has to do, rather, with *creativity*, the expression of truth and beauty, and also with *play*, what Hindu tradition terms *līlā*.

One final point needs to be made: nothing obliges us to suppose that a temporal process which is *not* productive of design, or of CSI, can *ipso facto* be attributed to natural causation. A case in point is given by the quantum-mechanical phenomenon of state vector collapse: the radioactive disintegration of a radium atom, for instance, cannot in fact be accounted for in terms of natural causation by virtue of its irreducible discontinuity. The ancient dictum "*Natura non facit saltum*," I claim, holds to this day; only it needs to be understood that the *natura* in question is *natura naturata* as distinguished from *natura naturans*: the "natured" as opposed to the "naturing." I have argued elsewhere that *natura naturata* acts invariably by way of a *continuous* temporal process, in contrast to *natura naturans*,

---

19. This does not mean that the intellective act occurs "before" the musical idea is conveyed to the world of sense: the intellective act has *ontological* rather than temporal priority. The act in question *can* in fact have no temporal priority, seeing that it is atemporal. It belongs to a realm where "before" and "after" do not apply, hard as this may be to grasp. To do so one requires in fact a symbol, a mental icon if you will: think of a circle, for instance, the circumference of which represents the realm of time, of "before" and "after." One sees that inasmuch as the center is equidistant to all points on the circumference, it is equally "present" to each, regardless of its position within the temporal sequence. That Center, one might add, is Dante's "pivot around which the primordial wheel revolves." There "every where and every when are focused" he tells us: "heaven and all nature hang" from that Point! (*Paradiso* 13.11, 29.12, 28.41, resp.)

which acts "above time" and thus by vertical causation.[20] The action of *natura naturans* is therefore inherently instantaneous; and it is this intrinsic instantaneity, I contend, that manifests in the phenomenon of irreducible discontinuity. The reason why state vector collapse has mystified physicists is quite simply that it cannot be attributed to natural causes. What disturbs the Einsteinians, it appears, is not merely the breakdown of determinism, but the insufficiency of natural causation: not only does God "play dice," but worst of all, He does so "instantaneously"!

The fact that the DNA in a living cell carries "tons of CSI" implies that Dembski's theorem has disqualified the claims of Darwinism. It has dashed the neo-Darwinist hope of finding "an algorithm, a natural law that leads to the origin of information" as Manfred Eigen has put it.[21] What stands at the very core of a living organism is CSI; and no natural law, no stochastic process, and no algorithm whatever can produce that CSI. No one can predict how long it may take the scientific community at large to accept this proven fact and draw the requisite consequences; certainly, if we admit what Thomas Kuhn has to say apropos of "scientific revolutions," this will not happen overnight. What in any case I find to be of far greater interest than the rise and fall of the Darwinist paradigm is the fact that Dembski's theory currently poses a fatal threat, not just to Darwinism, but to the cause of authentic religion, surprising as this may seem. To be sure, theologians for the most part are jubilant to learn that God is not superfluous after all; yet what they fail to realize, virtually to a man, is that the ID movement threatens to plunge us into an error worse than Darwinism. The problem is this: Whereas design theory has *de facto* disqualified the Darwinian mechanism, it has in no wise discredited the Darwinist concept of common descent, which thus remains entrenched as a scientistic dogma. But

20. *The Quantum Enigma,* op. cit., 105–108.

21. *Steps Towards Life: A Perspective on Evolution* (Oxford University Press, 1992), p. 12.

clearly, the hypothesis of a common descent that cannot be accounted for in terms of the Darwinian mechanism—nor, for that matter, in terms of natural causation as such—is tantamount to what has been termed "theistic evolution," a notion which brings God into the picture as a *deus ex machina* missioned to make Darwinian evolution work. It is virtually inevitable, therefore, that Dembski's discovery will be generally perceived as a scientific vindication of that tenet, a doctrine which has already swept the theological world and penetrated even into the Vatican. If theistic evolutionism has been the rage among theologians ever since the sixties when the writings of Teilhard de Chardin became the topic of the day, think what its status must be in the wake of Dembski's monumental discovery! Here indeed is a doctrine to "deceive even the elect."[22]

The problem with the tenet of common descent is that it obviates metaphysics in a domain which happens to be incurably metaphysical. Common descent—if there be such a thing—is something that transpires in space and answers to the demands of our common understanding. Therein, no doubt, lies its appeal. But therein, too, lies its impossibility: for as every traditional school has recognized, it happens that *first origins cannot in fact be situated in space and time*. There exists, in particular, a Patristic doctrine concerning first *biological* origins—I am referring to the doctrine of *ratione seminales* elaborated by St. Augustine in *De Genesis ad Litteram*—which proves to be incurably metaphysical. In particular, it alludes to a *vertical* descent, a progression from the metacosmic Center to the cosmic periphery, which constitutes an authentic *evolution*: an actual "unfolding" namely of a being which already exists. But this means that the scientist is evidently "coming in" near the end of the story: confined by the *modus operandi* of his approach to an exclusively horizontal perspective, he misses the vertical descent which ontologically precedes all manifestation on the spatio-temporal plane. In keeping with this horizontal perspective, the hypothesis of common descent proposes to resolve the mystery of first biological

22. I have dealt with this subject at length in *Theistic Evolution: The Teilhardian Heresy*, op. cit.

origins within the spatio-temporal domain: at the rim of the cosmic wheel, where indeed first origins *cannot* occur. Now, to bring God into the picture, as the theistic evolutionists have done, does not alter this fact, this principial impossibility: it only compounds bad science with heretical theology.

One forgets that authentic evolution is indeed an unfolding as we learn from the Latin verb *evolvere* (*e* + *volvere*, "to roll out"): an *outbound* kind of movement thus. Where there is an "outside," however, there must also be an "inside," an interior; and let me hasten to add that we must not psychologize this "inside": the *bona fide* interior of an organism is not a matter of "consciousness," but constitutes the ground, rather, from which every component of the organism, including consciousness itself, is derived. The integral organism—like the integral cosmos—may thus be itself conceived in terms of a symbolic circle, whose center represents its "innermost" point, the true *ratio seminale*,[23] of which the visible creature, situated in space and time, is but the outward manifestation, the term of an "unfolding."

It is moreover of interest to note that the Greek equivalent of the term *ratio seminale* is *logos spermatikos*: the "seed word," if you will, whose marred reflection, as I have suggested earlier, can indeed be discerned in the genetic code. The "origin of information," thus, which neo-Darwinists are seeking in an evolutionary algorithm, is actually to be found in the *logos spermatikos* that came into being in the single instant of creation, when "*God created the heaven and the earth.*" From that point of origin there began a "vertical descent," an *evolution* in the true sense of the term, in which however the essence of the organism, its *ratio seminale*, remains unchanged. Here, then, is the crux of the matter: to comprehend this metaphysical truth— even "*as through a glass, darkly*"—is to perceive at once the fallacy, not just of Darwinism, but of theistic evolutionism as well. But as I have indicated before, the latter doctrine is the worse of the two: for whereas Darwinism as such offends by an unwarranted extrapola-

---

23. According to Thomistic ontology, the *ratio seminale* of an organism is itself "external" to its act-of-being, which does in fact constitute its "innermost" point: the true center of the symbolic circle.

tion while remaining otherwise faithful to the scientific point of view, theistic evolutionism betrays the theological outlook itself, and in so doing engenders a wholesale corruption of sacred doctrine.

The point needs to be made, in particular, that where there is no *vertical descent*, there can be no *vertical ascent* either. A thing which has its first origin in space and time will have its last ending in space and time as well; such an entity is bound to perish, bound to disappear like a riven cloud. But such is not the case with things that have *being*, and thus an *essence* and an *act-of-being*. Only a cosmology, therefore, which enshrines the dimension of verticality can support a religious outlook and allow a doctrine of human immortality. Within the confines imposed by a horizontal cosmology, the claim of religion becomes perforce a sham, or at best a consoling fiction.

10

Interpreting
Anthropic Coincidence

What is a man, that the electron is mindful of him?
Carl Becker

By the middle of the twentieth century physicists had come to realize that the familiar world in which we live and have our physical being constitutes a highly improbable formation, a contingency about as unlikely as randomly picking a premarked grain of sand in the Sahara. By then it was clearly understood that a handful of fundamental numerical constants, such as the ratios of the nuclear, electric and gravitational forces, determine what is physically possible and what is not. One came to recognize in particular that the atomic and molecular structures upon which life depends require an immensely delicate balance or near-balance of opposing forces, which in turn demands that the constants of Nature assume the values they are found to have with almost a zero margin of tolerance: it is somewhat as if the distance between Earth and Moon, let us say, had to be what it is to within the thickness of a hair. In the face of such recognitions one can hardly refrain from asking *why* the universe should be thus fine tuned. Admittedly the question is more philosophical than scientific, given that the laws and constants of Nature constitute—by definition if you will—the principles in terms of which scientific explanations are framed. It is not, strictly speaking, the task of science to explain or justify the existence of these laws, or the values of these constants; and yet the fact remains that for more than half a century, scientists have debated the issue.

An entire literature, claiming at least to be scientific, has evolved in response to this problem.[1]

Although the fact of fine tuning is evidenced by the state of the universe here and now, it happens that the debate has from the start presupposed an evolutionist scenario. Two problems, in particular, have dominated the initial phase of the discussion. The first has to do with the fact that the very large integers and dimensionless ratios arising naturally from atomic physics and the new cosmology have turned out to be close to 10^{40} or its square. For example, the ratio N of the electric to the gravitational force between a proton and an electron is approximately 2.3×10^{39}. Compare this with another exceedingly large number of interest to cosmologists: the present age of the universe in so-called jiffies, a jiffy being the time it takes light to traverse the diameter of a proton. Now, this quantity, call it M, is calculated to be of the order 6×10^{39}. It was this close proximity between the seemingly unrelated quantities M and N that prompted Paul Dirac—one of the greatest physicists of the twentieth century, let us not forget—to propose, in 1937, what he termed the Large Number Hypothesis: "Any two of the very large dimensionless numbers occurring in Nature are connected by a simple mathematical relation, in which the coefficients are of the order of magnitude of unity." Since M is time-dependent, this implies that N must be time-dependent as well; and this fact led Dirac to propose a second hypothesis: he stipulated that the gravitational constant G is inversely proportional to the age of the universe. On this basis, M and N could remain comparable, and both would presently be large simply because the universe is old. And so the matter stood until 1957, when a Princeton University physicist by the name of Robert Dicke explained the numerical proximity of M and N in a very different way.

In keeping with the big bang scenario, Dicke assumed that the heavier elements upon which life on Earth depends, such as carbon and oxygen, were produced in the interior of a star through nucleosynthesis and released into the surrounding space when the star,

---

1. See especially the now classic treatise by John D. Barrow and Frank J. Tipler, *The Anthropic Cosmological Principle* (Oxford University Press, 1986).

having burned up its fuel, turned into a supernova and disintegrated. Now, this process, from start to finish, requires about 10 billion years. Allowing perhaps another few billion for the formation of planet Earth and the evolution of life, Dicke concluded that it takes somewhere between 10 and 15 billion years for scientific observers to appear on the scene. Taking into account estimated survival times, he inferred that "The age of the universe 'now' is not random but is conditioned by biological factors." This establishes a close connection between the lifetime of main sequence stars and the constant M representing the present age of the universe. A third large number L, in the form of a stellar lifetime, has now entered the picture; but it happens that L is related to N given that stellar evolution is determined by the interplay of electric and gravitational forces. Thus, by way of the constant L, Dicke was able to explain, at least to the satisfaction of his fellow scientists, a large number coincidence that had puzzled the astrophysics community for decades. It is to be noted that his explanation brings into play a novel principle, one that came in 1974 to be dubbed "the anthropic principle." Basically, it affirms that the universe must have certain features, because if it did not, human beings would not be here to observe that universe. In Dicke's case, the point at issue was simply that "carbon is required to make physicists," to put it in his own words. It turns out that this rather obvious fact does enable physicists "thus made" to hypothesize a close relation between two very large and seemingly unconnected numbers.

The second problem to which I have alluded has to do with the production of carbon in the interior of a star. It evidently takes three helium 4 nuclei to produce one carbon 12 nucleus. However, since 3-particle collisions are exceedingly rare, one must suppose that the production takes place in two steps: first, two helium 4 nuclei collide to produce a beryllium 8 nucleus, which then collides with another helium 4 nucleus to produce carbon 12. Now, the feasibility of these fusions hinges upon favorable nuclear resonances. It turns out, as Fred Hoyle ascertained in 1954, that a carbon 12 nucleus *must have* a resonance level close to 7.7 MeV if nucleosynthesis of carbon 12 is to occur. The following year, while on sabbatical at Caltech, Hoyle prevailed upon his nuclear physics colleagues to test this prediction; and as it happens, a resonance level at 7.656 MeV was in fact detected. Hoyle's unspoken reasoning, one sees, was once again "backwards"—from the existence of carbon 12 to its stellar nucleosynthesis, and thus from the present to the past—and therefore belongs implicitly to the "anthropic" genre. To be sure, the story of nucleosynthesis, as it is told today, does not end with the production of carbon, and one finds that a number of other "anthropic coincidences" enter perforce into the production of nuclei "required to make a physicist."

However, as I have noted before, the necessity of fine tuning is evidenced, first of all, by the state of the universe here and now. Let us consider a very simple example. The charge of an electron and of a proton as determined experimentally are found to be equal and opposite; their ratio is consequently −1. What would happen, let us ask, if this ratio were ever so slightly changed? Given the atomic constitution of matter, one finds that an explosion would take place: the opposite charges of protons and electrons would no longer cancel out, implying that material objects would become electrically charged (positively or negatively, depending on which of the two elementary charges dominates). Not only, therefore, would objects repel each other, but each would repel itself internally, which is to say that it would explode. A change in the ratio −1 by as little as 1 part in 100 billion would in fact suffice to blow our world to bits in a split second. Now, this is only the simplest example, one that can he appreciated without any deeper knowledge of physics. Obviously other more refined "anthropic coincidences" are required to permit the *existence*—as distinguished from a surmised *formation* via nucleosynthesis—of nuclei, atoms and molecules, as well as of their organic and inorganic aggregates, and to assure the myriad physical and chemical properties of these aggregates upon which our life depends. And remarkably, once nuclei are in place two fundamental constants suffice in principle to control this multitude of phenomena: namely, the so-called fine structure constant, whose value is approximately .00729720, and the ratio of the proton to the electron mass, which is close to 1836.12. According to quantum theory, a universe thus fine tuned will permit "carbon-based physicists" not only to exist, but to construct the technological apparatus by means of which they are able to explore the universe.

⊕

Clearly, the discovery of anthropic coincidence calls for a shift in the scientific outlook. Scientists had realized for a very long time that living creatures are adapted to the cosmic ambience; but now the realization has dawned that the cosmos itself is adapted to living beings. The adaptation, one finds, is mutual. It now appears that the cosmos, for whatever reason, has been fine tuned for the reception of life, and that the very particles comprising the universe prove to be "mindful of man."

The first enunciation of the so-called Anthropic Principle, one sees in retrospect, had failed to grasp the critical point: "Our location in the universe," Brandon Carter specified back in 1974, "is necessarily privileged to the extent of being compatible with our existence as observers." One senses an effort to accommodate the new insight within the old worldview, an attempt, if you will, to circumvent the crux of the matter. Carter's allusion refers clearly to what had come to be known as the Copernican Principle: the notion, namely, that the cosmos, on average, is quite the same everywhere, and that there is nothing at all special, let alone "central," in our cosmic location. In Carter's original version, the Anthropic Principle can be seen as a complement or correction to the Copernican in recognition of the fact that the human observer requires a terrestrial environment fine tuned to his exceedingly fragile constitution. Pick a cosmic location at random, and almost with absolute certainty it will prove lethal to man.

Understandably, other versions of the Anthropic Principle were soon to follow, and there exists by now a considerable collection of such formulations, representing various strands of thought. To be sure, all these—with the possible exception of the PAP to be discussed below—presuppose the evolutionist scenario, from star-formation to the production, once again, of physicists. The simplest version is the so-called Weak Anthropic Principle (WAP), which Barrow and Tipler formulate as follows: "The observed values of all physical and cosmological quantities are not equally probable, but take on values restricted by the requirement that there exist sites where carbon-based life can evolve, and by the requirement that the Universe be old enough for it to have already done so."[2] The connection with the early work of Dicke and Hoyle is unmistakable. Leaving aside the question what the term "probable" could possibly mean in this context, it is to be noted that Barrow and Tipler have elevated the hypothesis of cosmic and biological evolution to the status of an unquestionable dogma. Yet after enunciating their WAP, the authors immediately go on to say: "Again we should stress that this statement is in no way speculative or controversial. It expresses only the fact that those properties of the Universe we are able to discern are self-selected by the fact that they must be consistent with our evolution and present existence." There is thus nothing "speculative and controversial," supposedly, in the stipulation that life on Earth has evolved from nuclei produced out of hydrogen and helium in the interior of stars! Apparently this tenet is no longer a hypothesis, a theory in progress, but has become a dogma which can serve as the bedrock for further scientific discovery.

What actually proves however to be essential in the WAP does not hinge upon any scenario of cosmogenesis, but simply boils down to the obvious recognition that the observed universe must be compatible with the fact that it *is* observed. Thus reduced to its essential idea, the WAP is indeed nonspeculative and noncontroversial; and yet, so far from constituting a mere tautology, it entails consequences of great scientific interest. As a self-selection principle, it has been in scientific use for centuries: Copernicus, for example, applied this very principle when he explained the retrograde motion of planets as a self-selection effect resulting from the orbital motion of the Earth. The fact that it is *we* who are observing the universe—and not some disembodied spirit—does carry weight. It is this principle, precisely, that empowered Dicke to conclude that the observed age of the universe is more than "a brute fact," and it is rigorously true that within the context of the cosmogenetic

2. Op. cit., p.16.

scenario, that recognition does lead to an understanding of the large number coincidence he set out to explain.

Yet it is likewise clear that the WAP does not give physicists all that they want. Thus, when Hoyle made his famous prediction, he did so on grounds not ostensibly covered by that principle; the location of a nuclear resonance, after all, is determined by the strength of the nuclear force, and it is difficult to imagine that the fundamental constants of Nature could be affected by a scientific observer, one who, in fact, only came into existence some 15 billion years *after* the constants in question became operative. Clearly, the logical principle implicit in Hoyle's reasoning was something *more* than the WAP, something that has in fact come to be termed the Strong Anthropic Principle (SAP). As formulated by Barrow and Tipler, it affirms that "The Universe must have those properties which allow life to develop within it at some stage of its history." The evolutionist hypothesis has thus been incorporated once again into the formulation; but now, as an explicit component of the SAP, that hypothesis proves to be problematic even to the contemporary scientific mind.

It is a fact that the constants of Nature are what they are, and that life exists on Earth; what more, then, does the SAP demand? Is there a difference between "is" and "must be"? It would seem that a determination has somehow been made, a selection in favor of fundamental constants permissive of life. Such a fine tuning, however, at the start of a presumed cosmogenesis, is strongly suggestive of intelligent design, and seems to demand the postulate of a Creator. It was to be expected, therefore, that the enunciation of the Strong Anthropic Principle would lead to controversy within the scientific community, and cause divisions within its ranks. A certain minority, comprised of scientists with at least a minimal affinity for theological thought, accepted the SAP at its face value so to speak; for them the principle assumed the following form: "There exists one possible universe 'designed' with the goal of generating and sustaining 'observers.'"[3] Most scientists, on the other hand, having embraced the philosophy of naturalism, have found the notion of a

3. Ibid., p. 32.

"designed" universe—"designed" even for the purpose of ensuring their own existence!—quite unacceptable. The major segment of the scientific community found itself, thus, confronted by a problem of no mean proportions: how does one explain, in naturalist terms, why the universe *must be* so constituted as to generate and sustain "observers" at some stage of its evolution? Yet, daunting as the problem may be, one should never underestimate the resourcefulness of physicists: if it turns out—as it does in this case—that a single universe will not do, one overcomes this difficulty by postulating that there are many, and if need be, an infinite number. One can then speak of "probabilities" with reference to fundamental constants, and explain anthropic coincidence by an anthropic concept of self-selection. This option, which apparently has been chosen by a majority of scientists concerned with the anthropic problem, gives rise to yet another version of the SAP, which Barrow and Tipler formulate as follows: "An ensemble of other different universes is necessary for the existence of our Universe."

It should be mentioned that multiverse theory was first formulated to resolve another issue troublesome to physicists: the enigma of Schrödinger's cat, basically. Thus it occurred to a quantum theorist named Hugh Everett in 1957 that one can obviate the quandaries associated with measurement by introducing the notion of "parallel universes." Instead of admitting that a measurement "selects" a so-called eigenstate from a superposition of states—that it "collapses the state vector" as the expression goes—Everett proposed that it actually realizes *every* possibility by "splitting" the universe into as many parallel copies as there are states. In the case of Schrödinger's cat this means that the observer, by "opening the hatch," splits the universe—not "in half"—but "in two": one universe in which the cat is alive, and one in which it is dead. Of course this entails the equally bizarre notion that the observer himself is split in two: one finding the cat hale, the other that it has succumbed. But perhaps the most amazing thing of all is the fact Everett's idea has been taken seriously in the physics community, that it has in fact been championed by a highly respected coterie of "multiverse" experts. The doctrine that "every quantum transition taking place on every star, in every remote corner of the universe, is splitting our local world on earth into myriads of copies of itself. . . ."[4] ranks today as a prime contender in the world of high physics.

Three different ways of generating universes have in fact been proposed; following the suggestive terminology of Karl Giberson,[5] I will refer to these as Recycling, Splitting, and Bubbling. The first kind of generation is said to take place when a big bang generated universe eventually collapses into a terminal singularity, generally referred to as the "big crunch," and emerges on the other side, figuratively speaking, as a Recycled Universe with duly altered fundamental properties. Needless to say, of course, there is no *bona fide* physics that takes us through that singularity. But be that as it may, continuing in this manner one obtains formally an indefinite series of universes, from which ours has conceivably been self-selected by the fact that we happen to exist in this particular member of the given ensemble. The second universe-generating scenario, termed Splitting, is the one first conceived Hugh Everett in the context of quantum theory to explain the so-called collapse of the state vector in the act of measurement. The third, finally, the idea of Bubbling, was inspired by quantum field theory, which has something to say concerning the production of virtual particles as random fluctuations in a vacuum. As Giberson describes the resultant conception: "Our universe is but one of many embedded in a larger 'meta-universe', out of which new universes are constantly erupting like bubbles from the bottom of a pan of boiling water."[6]

Such are the ways, briefly stated, by which physicists have conceived a "many-worlds" scenario to make naturalistic sense out of anthropic coincidence. The attraction of multiverse theory in this context derives from the fact that it brings the problem within range of the WAP and thereby eliminates the specter of intelligent design, which comes now to be replaced by a concept of self-selection. The connection with Darwinism, in particular, should not be

---

4. B.S. DeWitt, "The many-universes interpretation of quantum mechanics," *Foundations of Quantum Mechanics* (New York: Academic Press, 1971).

5. "The Anthropic Principle," *Journal of Interdisciplinary Studies*, vol. 9 (1997), pp. 63–90.

6. Op. cit., p. 74.

overlooked. As it happens, the many-worlds approach has been suggested independently by a Cambridge University biologist named Charles Pantin: "If we could know that our Universe was only one of an indefinite number with varying properties," he argued, "we could perhaps invoke a solution analogous to the principle of Natural Selection" to explain the occurrence of anthropic coincidence. According to the theoreticians, all these many or innumerable universes are equally real, equally significant or insignificant: it appears that in this egalitarian age universes have likewise been accorded "equal status"!

I will mention in passing that a third basic version of the SAP, known as the Participatory Anthropic Principle (PAP), has been proposed by the eminent physicist John Wheeler. It too is inspired by quantum theory, and is closely related to the many-worlds approach. Its distinguishing feature, however, resides in the fact that it assigns an astounding role to the scientific observer: so far from constituting a mere spectator, it is affirmed that he actually brings the universe into existence! One sees, as in fact Barrow and Tipler point out, that Wheeler's theory has a distinctively idealist cast: the pendulum has now swung from Newtonian and Einsteinian realism to a position not unlike that proposed by George Berkeley in opposition to the Newtonian. It corresponds, in any case, to a third possible interpretation of the SAP, which reads: "Observers are necessary to bring the Universe into being." One wonders, of course, how there could be human observers before there is a universe; but then, if it be conceivable that one can split the universe in two by peeking at a cat, why not this as well.

A fourth—and in a way the most interesting—version of the SAP, known as the Final Anthropic Principle (FAP), states that "Intelligent information-processing must come into existence in the Universe, and, once it comes into existence, it will never die out." The idea derives from a paper by the distinguished physicist Freeman Dyson which appeared in the *Reviews of Modern Physics* in 1979 under the title "Time without End: Physics and Biology in an Open Universe," and from his Darwin Lecture, entitled "Life in the Universe," delivered two years later at Cambridge University. Based upon concepts presented in these papers a new mathematical discipline, presently known as "physical eschatology," has come to birth. According to this theory, living creatures—and intelligent creatures especially—are in effect computers, made up of hardware plus software embodying a program, which is seen as the scientific equivalent of what formerly was termed the "soul." Barrow and Tipler make the point explicitly:

> The essence of the human being is not the body but the program which controls the body; we might even identify the program which controls the body with the religious notion of a *soul*, for both are defined to be non-material entities which are the essence of a human personality.[7]

The intent of eschatology physics is evidently to show that somehow "programs do not die," which is to say that the information-processing can perpetuate itself into future states of the universe in which conditions will no longer allow the present-day kind of biological hardware. We refer the interested reader to Barrow and Tipler's Oxford University treatise, in which the proposed "principle of immortality" is expounded.[8] What confronts us here is a branch of physics—or at least a discipline that claims to be such and exhibits the trappings of a mathematical science—which seems quite overtly to encroach upon theological turf: what is one to make of that?

Following this exceedingly brief survey of the contemporary literature pertaining to the Anthropic Principle, it behooves us to reflect upon these putatively scientific theories. Even when it comes to their so-called Final Anthropic Principle, Barrow and Tipler insist that it is still "a statement of physics";[9] but is it really such? The same doubt applies obviously to the various many-worlds theories, not to speak of John Wheeler's Participatory Anthropic Principle.

---

7. *The Anthropic Cosmological Principle*, op. cit., p. 659.

8. See also Frank Tipler's later book, entitled *The Physics of Immortality* (New York: Doubleday, 1994).

9. Op. cit., p. 23.

What is it, then, that has led to this proliferation of highly dubious theories? The answer is clear: it is evidently the fact that the discovery of anthropic coincidence was perceived by the scientific community as a threat to their naturalistic and evolutionist *Weltanschauung*. Why *should* the electron be "mindful of man"? Why should the fundamental constants of Nature be fine tuned to 1 part in so many billion in order that life may spring into existence in the course of time? Could this be simply an accident? Even though it is of course unclear in what sense an entire universe could be "accidental," one has a vague sense that somehow more than "accident," more than "chance" must be at play. But what could that be? Having rejected the theistic interpretation out of hand, the scientist is hard pressed to arrive at an explanation of his own. The Weak Anthropic Principle seems to be as far as one can go on a rigorous scientific basis; beyond this point *the very question* is no longer scientific, but turns perforce metaphysical. Even the doubt—the very sense of wonder aroused by the discovery of anthropic coincidence—proves to be in essence the kind from whence authentic philosophy is said to spring. Such wonder invites us to reflect deeply enough to exit "this narrow world." Yet it seems that few scientists have responded to this call, hampered as they tend to be by an assortment of scientistic beliefs. To be precise, it is scientistic belief in the form of an evolutionist cosmology—and *not* "hard" physics—that has created the impasse "anthropic coincidence" theoreticians are attempting to resolve. The die had thus been cast: under these auspices—these scientistic assumptions!—there *is* no way out but to engage in some kind of science fiction assuming pseudo-philosophical forms.

More than just "hard" physics, thus, had come into play even before the advent of anthropic theorizing. The transition from big bang cosmology in its original form to Recycled, Splitting, and Bubbling Universes is by no means as radical or discontinuous as one might think, and it is in fact hard to determine where exactly the genre of authentic physics was first breached. One thing however is clear: by the time we arrive at a Recycled, Splitting, or Bubbling Universe, the boundary has indeed been crossed. We must remember that these "other worlds" are categorically unobservable, which is to say that there can be no causal connection between any object in such a world and our own: no signal or causal chain originating in one can ever reach the other.

Yet it could also be argued that the new disciplines constitute genuine physics by virtue of the fact that they were initiated by physicists in good standing, which include some highly respected names: this is by no means the work of amateurs or lunatics! The level of mathematical expertise and sophistication exemplified in such theories is typically high, and their connection with such reputable domains as particle physics and quantum field theory can be exceedingly close. Science fiction in the popular sense it is certainly not. A better designation for the emerging genre might be "hyperphysics": a physics, namely, which has transgressed its own proper bounds. One has the impression that physics itself, driven by the genius of its practitioners, is evolving into something new, something quite different from the "hard" science it used to be. This has led to an unprecedented proliferation of novel mathematical structures replete with "entities" the empirical basis of which is becoming ever more tenuous, while at the same time the levels of mathematical abstraction and complexity are reaching unprecedented heights. It is almost as if physics wanted to turn itself into a kind of mathematical metaphysics—a theology no less!—by virtue of its mathematical prowess and surpassing powers of abstraction. It is highly significant, in this regard, that Barrow and Tipler speak of their FAP as "the physical precondition for moral values"; and again I want to stress that this is something more than simple lunacy. One is in fact reminded of the Teilhardian vision of "science turned religion": is this not *precisely* what stands here at issue?

What confronts us is an endeavor on the part of science to exceed itself, to transgress its own proper domain, its own inviolable bounds, and thus turn itself literally into a "theory of everything."[10] One fails to realize that the very methodology of science imposes certain limits to which its findings must conform: that "hard" science in our day has in fact brought to light a number of what I term

---

10. The point has been made quite explicitly by Stephen Hawking in his latest book, *The Grand Design*. For a concise refutation of his views I refer to Chapter 7 in *Science and Myth*, op. cit.

limit theorems, which actually disclose certain of these bounds by the methods of science itself.[11] Who then can doubt that physics, by its very nature, is confined to this world in which we find ourselves: that in fact it refers perforce to a restricted domain thereof. As Goethe wisely observed: "*In der Beschränkung zeigt sich der Meister.*"[12] The very rigor and exactitude of physics derive in fact from its *Beschränkung*, its subjection to bounds. One might go so far as to maintain with Arthur Eddington and Roy Frieden[13] that the things with which physics is concerned are in fact *constituted* by the very bounds imposed upon our world through the *modus operandi* of physics itself; and let us note that this claim in itself entails a limit theorem to the effect that there *can be no* multiverse physics! In a word, physics can function precisely because it is *not* a "theory of everything."

Of course it is always possible to transgress bounds "on paper," and indeed one may do so with a semblance of rigor that commands respect; yet needless to say, the resultant structures can be no more than figments of a highly schooled and possibly brilliant imagination. A science, let it be said, that would cast off its yoke of limitation turns into pseudoscience in a trice. It may indeed retain the outer form of science, its seeming precision, but not its substance, never its quarry of truth. The structures of such a putative science—no matter how mathematically rigorous, sophisticated, or exacting its argument may be—prove in the end to be empty: even the celebrated criterion of "mathematical elegance" is powerless to bestow reality. One might remark in passing that this is presumably the reason why Albert Einstein produced rather little of lasting interest after 1917, the year in which he published his general theory of relativity, following which he appears to have relinquished the method of thought experiments and physical reasoning which had served him so well during the productive period of his life (roughly

---

11. Gödel's famous "Incompleteness Theorem," for instance, or the Heisenberg Uncertainty Principle, are cases in point. See my article "Science and the Restoration of Culture" in *Modern Age*, vol. 43, no. 1 (2001), pp. 85–93.

12. "In delimitation the master shows himself": this rather ill-sounding rendition is the best I can do.

13. See Chapter 2.

from 1904 to 1917). Mathematical physics at large seems now to have entered a similar phase in which the emphasis has shifted from the fruitful interplay of theory and experiment to a genre of pure theory, a kind of mathematical universe building which outstrips the experimental findings by billions of light years not to speak of universes other than our own. Perhaps, in this postmodernist age, the physicist not unlike the duly enlightened philosopher has come to regard truth as little more than a mathematical convention conforming to the demands of his intellect.

Having reflected at some length upon the mainstream response to the enigma of anthropic coincidence, let us now consider the theistic interpretation voiced by scientists with some measure of religious belief, which likewise has assumed a variety of forms. Not every author included within this category is in fact a religious believer. There is the example of George Ellis, a former associate of Stephen Hawking and presently a Professor of Cosmic Physics in Trieste, who espouses what might be termed the minimalist position within the theistic category. Ellis speaks cautiously of "a wider framework" of explanation, and concedes that there may be "an underlying structure of meaning beneath the surface appearances of reality, most easily comprehended in terms of deliberate Design," a surmise that might well be described as a first step in the right direction. So too the well-known physicist Paul Davies alludes to "an overwhelming impression of design" in a well-received book entitled *Cosmic Blueprint* (1988), and four years later, in *The Mind of God,* goes on to explain his own status:

> I belong to the group of scientists who do not subscribe to a conventional religion but nevertheless deny that the universe is a purposeless accident. Through my scientific work I have come to believe more and more strongly that the physical universe is put together with an ingenuity so astonishing that I cannot accept it merely as a brute fact. There must, it seems to me, be a deeper level of explanation.

The sentiment expressed in this remarkably forthright statement of belief is no doubt shared by many serious-minded scientists, and may be about as far as most are willing to go. A comparative few, on the other hand, do "subscribe to a conventional religion." Perhaps the most notable representative of this group is John Polkinghorne, the physicist who resigned a professorship at Cambridge University to become an Anglican priest. Polkinghorne finds the facts of anthropic coincidence to be consonant with the Christian teaching that God created the world. It is not, in his eyes, a matter of proving the existence of God, not an "argument" in the sense of Aquinas, but rather, as he says, a "consonance." To be sure, Polkinghorne accepts the evolutionist scenario—from big bang to the origin of species—without reservation or qualm, and it is this scenario, precisely, which he perceives to be "consonant" with Christian doctrine. Obviously this obliges him to reject the teaching comprised in the first chapters of Genesis, in keeping with the contemporary theological trend. His understanding of creation, he explains, is "ontological" as opposed to temporal; the biblical teaching, he maintains, "is not concerned with temporal origin, but with ontological origin. It answers the question: why do things exist at all? God is as much the Creator today as he was 15 billion years ago."[14] The God of Christianity, Polkinghorne insists, is not "the God of the edges, with a vested interest in beginnings," but indeed "the God of all times and all places." Now, this has certainly a ring of truth and would actually be Augustinian if it were not for the crucial fact that St. Augustine did not simply reject the historical interpretation of the hexaemeron—did not "demythologize" Holy Writ *à la* Bultmann—but *transcended* it rather from an authentically metaphysical point of view.

The rift widens, to say the least, when Polkinghorne goes on to explain that "It is in sustaining the fruitful process of the world that God is at work as the Creator." To be sure, "the fruitful process of the world" to which he alludes is none other than evolution as currently understood, which is to say that Polkinghorne is propounding the

14. "So Finely Tuned a Universe," *Commonweal*, August 1996, pp. 11–18. Unmarked quotations from Polkinghorne are taken from this article.

Teilhardian notion that "God creates by way of evolution." In keeping with presumed insights of contemporary science he perceives the "fruitfulness" in question as resulting from "an interplay between two opposing tendencies which we could describe as 'chance' and 'necessity.'" And he goes on to enunciate a theological interpretation of these two principles: "I believe that the Christian God, who is both loving and faithful, has given to his creation the twin gifts of independence and reliability, which find their reflection in the fruitful process of the universe through the interplay between happenstance and regularity, between chance and necessity." At this juncture a logical point needs to be made: If indeed God creates by sustaining "the fruitful process of the world" consisting of an interplay between "chance" and "necessity," then it would follow that "the twin gifts of independence and reliability" are not merely "reflected" in chance and necessity, but must in fact coincide with these complementary modes of causation. In other words, so far from constituting secondary principles reflecting a primary duality ("the twin gifts of independence and reliability"), chance and necessity are in fact conceived as constituting the primary duality itself. Thus, when Polkinghorne speaks of "reflection," he is at this moment falling back to a pre-Teilhardian concept of creation inconsistent with his own.

A typical example of Polkinghorne's "fruitful process" is furnished by the formation of stars and galaxies as currently conceived. An originally smooth universe becomes grainy or lumpy by chance, and such small initial deviations become subsequently amplified through the action of gravity, which is where necessity enters the picture. Polkinghorne concludes that "the interplay between those tendencies, chance as origin of novelty, and necessity as the sifter and preserver of the novelty thus produced, is the prime way in which the fruitfulness of the universe is realized." Let us not fail however to observe that this is quintessential Darwinism, now applied on a cosmic scale.

It needs above all to be pointed out that in the proposed synthesis of science and theology what fits the least is the Christian doctrine of the Fall together with the associated dogma of Original Sin. It will be recalled that Teilhard de Chardin himself had struggled with this problem, and that despite his phenomenal powers of imagination he did so with very little success. "The principal obstacle encountered by orthodox thinkers," he wrote, "when they try to accommodate the revealed historical picture of human origins to the present scientific evidence is the traditional notion of original sin."[15] Now, it seems that Polkinghorne is no exception to this rule. In *Reason and Reality*, for instance, he deals with this issue in the final chapter, comprising exactly five pages. He does not even attempt to square the Fall with the evolutionist scenario but seems intent rather to devote these five pages to an explanation why the doctrine proves problematic in light of scientific findings. "Genesis 3," we are told in the end, "is to be understood as a myth about human alienation from God and not as an aetiological explanation of the all too evident plight of humanity."[16] One might of course raise the question first posed by Teilhard de Chardin: "Is this still Christianity?"—but that is another matter. What concerns us at the moment is the fact that even when it comes to scientists perceived to be Christian believers, it appears that the presumed evidence of science carries greater weight by far than the traditional teachings of the Faith: wherever one encounters a seemingly irreconcilable conflict between these respective teachings, it is invariably the Christian dogma that is cast out.

Let us not fail to note, in particular, that the very notion of "sin," and indeed of *responsibility*, has no more place in an evolutionist universe where freedom must be read as "chance"; for clearly, in chance *there is no* responsibility, no moral good or evil. The mystery of human freedom—of what is traditionally referred to as "freedom of the will"—lies deeper by far than the notion of chance. It has to do, not with matter, but with the opposite ontologic pole, the one which has been left out of consideration by the Darwinists. True freedom, as I have argued in the preceding Chapter, enters the world by way of *vertical* causation: a mode of causality which acts

15. *Christianity and Evolution* (New York: Harcourt Brace Jovanovich, 1971), p. 36. For a thorough analysis and critique of the Teilhardian doctrine I refer to my monograph *Theistic Evolution: The Teilhardian Heresy*, op. cit.

16. *Reason and Reality* (Hauppauge, NY: Trinity Press International, 1991), p. 73.

above time, and for this reason can find no place in an evolutionist cosmology. Freedom thus is neither chance nor necessity, nor a combination of the two, but something that exceeds the plane of natural causation, and which consequently cannot be understood in contemporary scientific terms. But clearly, a theology without even the *concept* of human freedom and moral responsibility is not in truth a theology at all. Such a doctrine can in fact be no more than a poor imitation, a clumsy *Ersatz*.

Yet it happens that Polkinghorne's outlook is defective even from a scientific point of view. In light of intelligent design theory, one now knows that the twin principles of chance and necessity do *not* suffice to account for "the fruitful process of the world." It turns out, speaking in contemporary scientific terms, that "fruitfulness" translates into complex specified information or CSI, and this is precisely what chance and necessity, singly or in combination, cannot produce.[17] The most striking example, to be sure, is provided by the "tons" of CSI in the DNA of every living creature, from a bacterium or ameba to the human organism, not one of which could in truth be produced by a stochastic process. If, therefore, one assumes that existing species have evolved by way of common descent, one is obliged to posit acts of vertical causation to account for the production of the requisite CSI. Presumably these interventions from above will be conceived as more or less localized at the branch points of the stipulated genealogical tree, in keeping with the idea of "saltations" espoused by neo-Darwinists such as Richard Goldschmidt and Stephen Jay Gould. In one way or another, vertical causation *must* come into play, even from a strictly scientific point of view. But it appears that Polkinghorne was not privy to this fact when he first enunciated his theory, which only became known in 1998.[18] Little did he realize, when he lectured in 1996, that the quintessential Darwinism, which he not only accepted without question but elevated to theological status, would be rigorously disproved within two years.

17. Again I refer to the preceding Chapter.

18. The decisive breakthrough came with the publication of William A. Dembski's *The Design Inference* (Cambridge University Press, 1998).

Given that an act of vertical causation on a cosmic scale is tantamount to an act of God, it appears that the ancient "argument from design" does after all carry validity. With the mathematical discovery of what Bill Dembski refers to as "design inference," natural theology in the sense of William Paley has been revived. By way of a curious dialectic a materialist science, which seemingly had banished God centuries ago, has now arrived at an impasse which only that "useless hypothesis" can break. By this very token, however, one sees that the "God of evolution"—or better said, the God of theistic evolutionists—is perforce a "God of the gaps," a *deus ex machina* whose function it is to produce the requisite tons of CSI at critical junctures where the stochastic process of classical Darwinism fails. Yet the fact remains that the proposed marriage of science and theology proves to be injurious to either side: a matter or propping up *bad* science by means of *heretical* theology. It is somewhat as if a mathematician, finding that the two sides of an equation do not balance, were to add a "God term" to make up the difference: far better to go back to the drawing board and check one's calculations. In view of the fact that common descent across the board proves not to be possible, the correct scientific response would be to abandon the Darwinist theory as a failed hypothesis. Why on earth should a scientist invoke God to rescue a failed theory: what kind of science is that?

To be sure, a theistic reading of anthropic coincidence is by no means unjustified: the idea that the universe has been designed to serve as a habitat for man is after all traditional. The problem, however, with contemporary theistic interpretations of anthropic coincidence resides in the fact that these interpretations invariably presuppose the evolutionist scenario. The God alluded to is thus indeed the God of evolution, the *deus ex machina* who enables common descent. One role of this hypothetical God, evidently, is to fine tune the cosmos, to adjust the fundamental constants so that carbon 12, for example, will have a nuclear resonance at 7.656 MeV. This kind of "theology" appears to legitimize all manner of scientific claims by conveying the impression that these are literally "God-given" facts. Who can argue, say, with a quantum field theory put in place by God himself? The new collaboration between science and theology proves thus to be a boon to both sides: for the theologian it means that at long last he will once again be taken seriously, whereas for the scientist it entails that the prestige of God, no less, will rub off on his theories.

Whether we realize it or not, theology itself is undergoing a profound transformation: the new affiliation with science is having its effect. Brand new theologies are in fact coming into vogue. I remember attending a symposium at which a Catholic Archbishop, after delivering a lecture on the work of Stephen Hawking, confided during the ensuing discussion that he personally prefers "process theology" to Trinitarian doctrine. What surprised me the most was that not one individual among the assembled scientists, theologians and priests seemed to be in the least taken aback or even astonished by this amazing revelation. I sensed the kinship with Teilhard de Chardin, and came away with the distinct impression that his lifelong preoccupation—"the effort to establish in myself and to spread around a new religion (you may call it a better Christianity)"[19]—was bearing bountiful fruit. Whatever one may think of the Teilhardian doctrine, it must be admitted that his vision of a new religion "burgeoning in the heart of modern man from a seed sown by the idea of evolution"[20] was nothing less than prophetic.

So much for contemporary theorizing of whatever stripe. I propose now to examine the conundrum of anthropic coincidence from an authentically metaphysical point of view, based upon the recognition that the problem *is* metaphysical: incurably so! What has driven the discourse from the start, it turns out, are not in fact scientific findings, but metaphysical beliefs: *spurious* metaphysical assumptions no less. It is needful, therefore, to expose these hidden premises, bring them to light—and this is something only a veritable metaphysics can do.

The task proves actually to be simpler than one might imagine;

19. *Lettres à Léontine Zanta* (Paris: Desclée de Brouwer, 1965), p.127.
20. *Activation of Energy* (New York: Harcourt Brace Jovanovich, 1970), p.383.

for it happens that a single misbegotten premise, presupposed across the board in the anthropic debate, stands at the root of the problem. Certainly other metaphysical misconceptions enter the picture as well; yet to bring down the edifice of spurious theorizing it suffices to expose the foundational error upon which that conceptual structure rests. What then could that notion be? Now, it happens that the answer has been staring us in the face from the start: since Chapter 1 to be precise. The offending premise, it turns out, proves to be none other than the assumption that corporeal objects, be they animate or inanimate, are "made of atoms": that they reduce, in other words, to an aggregate of fundamental particles. It is this presumed reduction of the *corporeal* to the *physical* that I identified in *The Quantum Enigma* as the source of seeming quantum paradox: the reason why "no one understands quantum theory" to put it in Feynman's words; and now we encounter this beguiling reduction once again, this time at the base of anthropic theorizing in its multifarious forms.

The question which all such speculation is designed to answer is simply this: Why should the laws and constants of physics be "fine tuned" to permit an evolution culminating in the appearance of man? Now this query presupposes not only an historical order—from fundamental particles, namely, to ever more complex molecules, culminating in the formation of man—but an *ontological* order as well: a reduction, namely, not only of intelligent life to the corporeal domain, but first of all a reduction of that domain itself to the physical. The very question anthropic theorizing is designed to answer hinges thus upon a misconception: one assumes that corporeal being reduces to an aggregate of fundamental particles when in fact it does not.

The anthropic theoretician is actually putting the cart before the horse: he basically conceives the whole as the sum of its parts. But in reality the whole is *not* the sum of its parts: even quantum theory is founded upon that recognition! Actually it is the other way round: the whole, namely, comes first, and the parts are in fact defined as parts with reference to the whole. I will note in passing that according to the Aristotelian definition of *quantity* as "that which admits mutually external parts," the prevailing reduction would place the corporeal entity squarely within the category of quantity, a stipulation which no authentic metaphysics could ever admit. Such then is the premise upon which contemporary anthropic theorizing is based: the ontological assumption presupposed throughout the length and breadth of the anthropic debate by theists as well as naturalists, by proponents of design no less than by advocates of chance.

The first thing, thus, that needs evidently to be done by way of ontological rectification is to discard the aforesaid premise, the cherished notion that corporeal entities are "made of atoms," that they are finally "nothing but" an aggregate of quantum particles. Whatever may be the ontological status of the particulate aggregate SX associated with a corporeal object X, the fact is that X and SX belong to different ontological domains.[21] One must bear in mind, moreover, that the whole invariably has primacy in relation to the parts; and if it turns out that the latter are in some sense "fine tuned," this must be due to their formation as parts of that whole. It is thus neither chance nor divine intervention that "fine tunes" quantum particles, but corporeal being, the corporeal order as a whole—interacting presumably with the strategies of the physicist—which accomplishes that feat. The very idea that God adjusts the laws and constants of Nature—that he "monkeys with physics" as Fred Hoyle once put it—to produce the organic world turns out thus to be ill-conceived and backwards. What God does, theologically speaking, is to *create*; and what He creates—*ex nihilo* and *omnia simul* as theology teaches—is not an assembly of quantum particles, but *a cosmos comprised of beings*. Particles come later, ontologically speaking, which is to say that they constitute a secondary reality. And as I have argued at length in Chapter 2, it is in a sense man, and not God, who "makes" these particles. John Wheeler was right: we do find ourselves in a "participatory universe"; only it needs to be understood that the universe to which Wheeler refers is the *physical* as distinguished from the *corporeal*, the existence of which he seems not to have recognized.

One sees that the picture has changed drastically, has in fact become inverted. To understand the formal logic of this inversion

21. See *The Quantum Enigma*, op. cit.

let us consider the following purely geometric scenario. Imagine a 2-dimensional "universe" in the form of a trapezoid ABCD with base AD, and suppose that astrophysicists residing in this universe have arrived at a big bang cosmology by the following means. Having ascertained the length of the sides and the interior angles of the trapezoid, and having calculated the coordinates of the point O at which the lines determined by the segments AB and CD intersect, they hypothesize that ABCD must have originated as a rectilinear emanation from that point of origin. And lo: an anthropic coincidence has now sprung into view! For it happens that the angle φ subtended by OB and OC is precisely what it needs to be to give rise to the measurable properties of ABCD: "If the initial angle of dispersion had differed from φ by so much as a billionth of a degree, *we* would not exist." Now, to the resident naturalists this suggests that there must be an infinite number of parallel universes, presumably in the form of trapezoids, corresponding to all possible values of that initial angular dispersion, from which the present universe has been self-selected in accordance with the WAP; to others perhaps it signifies that φ was determined from the start by an act of intelligent design. It turns out, however, that the triangle OBC corresponds precisely to what I have termed "the extrapolated universe,"[22] and that φ is actually determined neither by chance nor by providence, but simply by the trapezoid ABCD. It comes down to the following principle, which proves to be self-evident: *The actual universe determines the extrapolated universe, constants and all.* What the anthropic theoreticians have done has now become clear: *they have inverted the causal nexus.*

It needs however to be admitted that the evolutionist contention is not easily refuted on a scientific plane, and that a number of puzzling questions remain for which no ready answer appears to be at hand. Consider Fred Hoyle's prediction of the celebrated nuclear resonance in carbon 12, for example. Given that carbon 12 nuclei can only be produced by way of nucleosynthesis in the interior of stars, Hoyle had argued that whatever nuclear resonances are required to permit such a synthesis must in fact be there; and his

22. See Chapter 5.

prediction has proved to be correct. Now, one knows that nuclear resonances are determined by the fundamental laws and constants of physics; and from a non-evolutionist point of view there is no apparent reason to suppose that the relevant constants should be fine tuned so as to permit the whole series of nuclear reactions by which the elements required by living organisms can be built up, step by step, out of hydrogen and helium. I say "apparent" reason, because it is quite conceivable that as our knowledge deepens, a physical explanation of this fact may yet come to light: the history of science, after all, provides countless examples of such unforeseen and virtually unforeseeable discoveries. At present, however, we have no such explanation for the nuclear resonances in question: we have only Fred Hoyle's anthropic argument, which is based upon the evolutionist hypothesis.[23] The verification of Hoyle's conjecture counts thus as evidence in support of that hypothesis. As things stand there is scientific evidence both for and against the current status quo, and it will likely depend upon the ideological orientation of the individual scientist which way he tilts.

My point, however, is that the question *cannot* in fact be resolved on what counts today as scientific ground. As we have seen, it is first of all needful to jettison the metaphysical assumptions routinely presupposed by scientists, beginning with the Cartesian postulate of bifurcation. It then becomes evident that the scientist is attempting to solve the mystery of cosmogenesis where in fact it cannot possibly be resolved: on the physical plane, namely, which metaphysically speaking is *sub-existential*. As we have noted in the very first chapter, the physicist is unable in principle to comprehend corporeal being, unable to understand what transpires even in the simplest act of measurement: what to speak of living organisms!

Let us return, then, to our metaphysical reflections, which have not yet run their course. Certainly it is needful to jettison the bifurcation postulate, the misconceived reduction of the corporeal to the physical. Yet the fact remains that organic substances do decompose

---

23. What stands at issue here is primarily the physical component of the evolutionist hypothesis, which amounts to big bang theory. We have touched upon the evidential basis of that theory (apart from the anthropic issue) in Chapter 6.

into such physical entities as protein molecules, which implies that a fine tuning of physical constants is still in evidence, just as before. What has drastically changed, however, is the underlying perspective, which now gives primacy to the corporeal in relation to the physical. And this entails that the rationale for the fine tuning resides, not in the physical, but in the corporeal domain: it *must*, by reason of ontological precedence. Another simple-minded example may help to make this clear. If one breaks a clay pot one finds that the resultant shards fit together perfectly so as to constitute the erstwhile pot; and obviously this "fine tuning"—which seems almost miraculous so long as one does not know the true provenance of the shards—is the result neither of chance nor of design. And let us note that to reach this conclusion we need not know whether it was a hammer blow or perchance an accidental fall onto a marble floor that smashed the pot, much less such details as the angle of the blow or the speed at which the falling pot may have struck the ground. It suffices to know the provenance of the chards: the fact that they derive from the given pot. So too corporeal primacy as such suffices in principle to explain why the physical universe proves to be "fine tuned."

It happens however that more can be said: for as we have noted in Chapter 2, there is now strong evidence in support of the contention that the laws and fundamental constants of physics are in fact determined by the process of measurement as Eddington had foretold: it appears that Roy Frieden's information-theoretic analysis actually proves as much.[24] It is impossible, therefore, that the constants of physics should *not* be fine tuned so as to allow the economy of measurement, and be thus permissive of physicists. The underlying logic here is quite the same as in the scenario of the broken pot.

But the question remains: how does man, the *anthropos,* enter the picture? If the Anthropic Principle has to do simply with the fact of corporeal primacy, one should drop that designation and speak instead of a Corporeal Principle. We need now however to take a closer look at the process of measurement: in the first place it is to

---

24. *Physics from Fisher Information* (Cambridge University Press, 1998).

be noted that a scientific instrument is more by far than mere corporeal matter, which is to say that it constitutes an engineered artifact. Something absolutely essential, obviously, has been added by man. It is thus the scientific observer himself who contributes the complex specified information or CSI to which Nature responds in the act of mensuration; and as Roy Frieden has demonstrated, it is precisely this interaction that determines the laws and universal constants of physics. But not without the intervention of another factor: "Whatever we have to apprehend," says Eddington, "must be apprehended in a way for which our intellectual equipment has made provision."[25] It is thus that the *anthropos* enters the picture a second time: here, on this distinctly "anthropic" level, a process of selection and formation takes place which proves to be essential. In terms of Eddington's metaphor of the "net" one can say that the latter has both a *corporeal* and a *conceptual* component. The physicist functions thus in the fullness of his trichotomous nature, much like the artist who works externally with chisel or brush, and internally "through a word conceived in his intellect" as St. Thomas Aquinas informs us.[26] Here too, in the *modus operandi* of physics, a "word"—emanating from the Logos—comes into play.[27] It needs however to be understood that the scientist does not act by himself, but "participates" rather in the primary Logos, "*the Word that was in the beginning*": truly, "*Without Him was not made anything that was made.*"[28] Even secondary realities, it turns out, cannot be "*made without Him*": for it is precisely by virtue of the aforesaid "participation" that the scientist derives the prowess to create what Eddington terms the "intellectual equipment" needed to bring secondary realities into existence. As we have come to see, the physicist too is a "maker," an artist of a kind; and like every *bona fide* artist, he too acts "in imitation of *natura naturans*," the primal Artist who is none other than God.[29]

Now, this insight sheds light on many things. It explains, in par-

---

25. *The Philosophy of Physical Science* (Cambridge University Press, 1939), p. 115.
26. *Summa Theologiae* I, Quest. 117, Art. 1.
27. On this question I refer back to Chapter 9, especially pp. 195–201.
28. John 1:1, 3.

ticular, what Albert Einstein has termed "the most incomprehensible thing about the universe": the fact, namely, that it proves to be comprehensible. One sees now that the physical universe is understandable to the scientist for basically the same reason that a work of art is meaningful to the artist; it turns out, moreover, to be *mathematical* because it was in a sense "made" by mathematical physicists: "The mathematics is not there till we put it there" declared Eddington to the amazement of his colleagues.[30] "We have discovered a footprint in the sand" he exclaims, "and lo! it is our own." Here at last we have it: in this decisive recognition lies the ultimate significance of the Anthropic Principle, its burden of truth.

29. I am alluding specifically to the Scholastic doctrine of art; see especially Ananda Coomaraswamy, *Christian and Oriental Philosophy of Art* (New York: Dover, 1956).

30. Op. cit., p. 137.

11

Science and the
Restoration of Culture

My first point is scarcely controversial: From the eighteenth century onwards, I maintain, science has been the major determinant of culture in the West. The influence may be direct or mediated, and the response affirmative or oppositional, but the fact remains that in every cultural domain science has played a pivotal role as the prime agent of change. Take philosophy or theology, social or political norms, art, morals or religious practice: the story is the same. Like it or not, science is the decisive factor—the great new revelation—to which society at large has for long been reacting in multiple ways. Even as technology, the offspring and partner of science, has radically transformed the outer life of Western civilization, science itself is having its impact upon our inner life: upon our basic beliefs, values and aspirations. Not everyone, of course, has become an outright materialist; but all, I submit, have been profoundly affected nonetheless.

From its inception the new science has prospered visibly, and commended itself within ever-widening circles as the great liberator from ignorance and superstition. The age of Enlightenment was upon us, and in a very real sense, still is. Was not Bertrand Russell speaking for the modern world as such when he declared: "What science cannot tell us, mankind cannot know"? An exclusive faith in science appears indeed to be the hallmark of modernity.

That faith itself, however, has begun to falter: we have entered the era of postmodernism. It is not simply a matter of one worldview triumphing over another, as has happened in the past. The shift to postmodernism is far more radical than that; for it denies the validity, not just of an antecedent worldview, but of worldviews in general. Truth has been reduced in effect to a social convention, the local construct of a society. Partly in reaction, no doubt, to the tyranny of the scientistic *Weltanschauung*, one has set about to relativize *all* worldviews. What confronts us here, moreover, is not simply a philosophic trend, but a cultural phenomenon: a cultural revolution, one can say. Think of the wholesale rejection of traditional norms, the pervasive distrust of authority, the radical disorientation which seems especially to afflict the youth of our day. There are of course notable exceptions and indeed counter-trends; but these do not offset the nihilistic tendencies in questions. One has reason to believe, moreover, that there is a real connection between postmodernist philosophy and corresponding cultural trends, even if it may not be possible to construe that connection as a simple case of cause and effect. One can therefore speak of postmodernism in a broad sense, which includes its cultural manifestations.

What I wish now to point out is that postmodernism is not simply an oppositional reaction to the antecedent modernism, but is in fact implicit in modernity, that is to say, in the scientistic worldview itself. The universe as depicted by modern science is clearly unacceptable as a human habitat. The scientistic *Weltanschauung* is bearable, thus, precisely because no one believes it—I mean, fully, with all his being. We believe in the scientistic universe with a part of our mind, persuaded that the contention has been validated by rigorous scientific means; and yet we still suppose, in our daily lives, that the grass is green and the sky blue (which scientism denies), not to speak of the fact that we take a man or a woman to be more after all than a "chemical machine." We have learned to compartmentalize our beliefs: to pass in a trice from one persuasion to another, incompatible with the first, and think nothing of it. This way of managing beliefs needs of course to be learned; it is what modern education has done for us. The art is acquired in schools and universities. The practice, to be sure, is astonishing, if only one stops to think of it; but we generally don't. We have learned the art so well that we are hardly conscious of doing anything at all. As is the case in schizophrenia, we are unaware of our own inconsistency, until of course we engage in authentic philosophical reflection; but even then we rarely perceive the magnitude of the dilemma. It takes a Kierkegaard or a Nietzsche, apparently, to become profoundly disturbed. For most of us the anguish is potential rather than actual, it seems.

It appears from these sparse indications that postmodernism is latent in the scientific mentality. To oscillate between two contradictory worldviews is to commit to neither: to commit to nothing at all. As a chronic condition the practice is tantamount to a denial of truth.

Given that the scientistic outlook is humanly untenable, it behooves us to ask whether the *Weltanschauung* in question is essential to science as such. As one knows, modern science began as an amalgam of Cartesian metaphysics and Baconian empiricism, the incongruity of which was spotted soon enough by leading philosophers. The union, it turns out, is not a true synthesis, and what matters, in fact, is not the Cartesian ontology, nor its epistemology, but precisely the Baconian method. It is Bacon's *novum organum*, his "new machine for the mind," that enables the enterprise of modern science, a science in which "human knowledge and human power meet in one" as Bacon had foretold. To be sure, the Cartesian conception of *res extensa* (of "bare matter") has played a vital role in the motivation and guidance of scientific inquiry. As a Kuhnian paradigm, however, the notion of a clockwork universe is expendable; it is not an essential of science, but only a transitional aid. What ultimately counts, I say, is the methodology, the Baconian character of the enterprise.

The primary reductionism of science is thus methodological; it applies, not to reality as such, but to the means by which we propose to grasp and harness reality. Directed as it is to the objective of control, the Baconian enterprise is inherently designed to count, measure, and quantify; nothing in fact fulfills the Baconian guidelines more perfectly than a mathematical physics. This methodological reductionism, however, does not presuppose, nor entail, an ontology; it is metaphysically neutral, one can say. But whereas science does not *de jure* authorize a reductionism of the ontologic kind, it does so *de facto*; as a rule the tendency to deny what science cannot grasp proves irresistible. It is a fact that science begets scientism, and bifurcation is doubtless the primary scientistic dogma. As Gilbert Durand has wisely observed: "Dualism is the great 'schizomorphic' structure of Western intelligence."

It goes without saying that this "structure" of Western intelligence implicates a worldview and indeed a culture. I concur moreover with Theodore Roszak that "there are never two cultures; only one—though that one culture may be schizoid." And such indeed is our predicament. It should of course be added that not everyone living within the ambience of that culture is *of* that culture—but this is another matter. For better or for worse, there *is* a Western culture, even as there is a Western worldview; and both derive their support from Western science. Certainly that support is illegitimate; yet in terms of effectiveness, this fact carries no weight at all.

Having spoken of bifurcation as the primary scientistic dogma, I should point out that there exists a plethora of secondary scientistic dogmas which hinge upon the primary. Take Darwinism, for example: in response to those who think of that doctrine as a well-substantiated scientific theory I will refer to the growing scientific literature which proves that it is not.[1] Darwinism, it turns out, has never met the Baconian criteria of scientific legitimacy. However, given bifurcation—plus the associated idea that the universe consists of atoms or fundamental particles moving to no purpose, whether by chance or in accordance with deterministic laws—given this reductionist scenario, I say, there *is* basically no other way of conceiving biogenesis and speciation. And this, when all is said and done, is the principal reason why scientists continue to cling to some form of Darwinism despite its astronomical improbability. Darwinism, thus, is finally to be ranked as a scientistic dogma. But whereas this particular dogma constitutes evidently a prime example of scientistic belief, our textbooks are filled with tenets no less

---

1. For instance, Michael Denton, *Evolution: A Theory in Crisis* (Bethesda, Md.: Adler & Adler, 1986); Phillip E. Johnson, *Darwin on Trial* (Downers Grove, IL.: InterVarsity, 1993); Michael J. Behe, *Darwin's Black Box* (New York: The Free Press, 1996); William A. Dembski, *The Design Revolution* (Downers Grove, IL: InterVarsity, 2004); and Stephen C. Meyer, *Darwin's Doubt* (New York, NY: Harper Collins, 2014).

spurious, which likewise claim the status of scientific truth. It suffices that these tenets fit into the prevailing worldview and lend a kind of mutual support; the fact that they do not pass scientific muster remains generally unrecognized, and perhaps would not cause too much consternation if it did become known. It is only when a crisis arises in a particular field that scientists—some scientists, at least—become motivated to engage in foundational inquiries; and even then, it appears, a clear-cut discernment between scientific fact and scientistic fiction is rarely achieved.

One sees that assumptions of a philosophic nature, as well as ideological commitments, do affect the scientific enterprise, which in fact is not quite as "scientific" as one tends to suppose. Scientists are human, after all, not robots or computers; and the postmodernist philosophy of science does after all have a point. And yet, surprisingly perhaps, there *is* such a thing as "hard" science: a rigorous discipline capable of real discovery. Such science carries its own exactitude which no man can bend, and discloses objects or theorems which—like Mount Everest—are simply *there*. Hard science, it turns out, is wiser in certain ways than the scientists who engage in its pursuit: with a single decree it can abolish a long-standing expectation or disqualify some hallowed canon of scientific belief; it has in a very real sense a life of its own, independent of social conventions, philosophic bias, or ideological orientation. Apart from technical competence and occasional genius, it demands just one thing from the scientific community: integrity, namely, a certain respect for truth. And happily it can be said, to the honor of that community, that its members have by and large proved worthy of this trust.

It needs now to be pointed out that something momentous and utterly unexpected has taken place within the scientific domain in the course of the twentieth century: science has begun at last to discern its own inherent limitations, its own categorical bounds. Not that it has disavowed its exactitude: not at all! What science *has* disavowed is the scientistic notion that these exactitudes apply in principle to every domain, that *de jure* science encompasses all truth. It has moreover arrived at this recognition of its own incapacity, not by way of some supra-scientific intuition, but by strictly scientific means. What stands at issue are indeed theorems, discoveries as inexorable as the certitudes of mathematics or the fundamental laws of Nature. I propose now to cite a few major examples of such twentieth-century "limit theorems," spanning the gamut from mathematics and physics to biology and cognitive psychology; the cultural implications of these discoveries will occupy us later. For the moment it suffices to note that these remarkable findings are supportive of my "absolutist" claims in behalf of what I have termed *hard* science.

I will cite, as my first example, the Incompleteness Theorem established in 1931 by Kurt Gödel, a 25-year-old Austrian mathematician, which arguably constitutes the most important discovery of a logical kind in the twentieth century. Gödel's theorem disqualifies, at one stroke, the long-held expectations of leading authorities, the likes of David Hilbert, Gottlob Frege and Bertrand Russell, who thought that a formal system inclusive of all mathematical truth could be found. What the young Austrian proved— once and for all!—is that a consistent formal system rich enough to accommo-date ordinary arithmetic is necessarily incomplete; there simply *is* no formal structure encompassing all mathematical truth. It may be noted that Gödel's theorem has a certain postmodernist ring: by restricting the scope of a single theory, a single formal system, it seemingly opens the door to a pluralist outlook tolerant of alterna-tive positions. But even so, it does not compromise the absolute claims of truth: Gödel's result, after all, is a theorem of mathemati-cal logic, validated by a rigorous argument, an incontrovertible proof. It does not in any way relativize mathematical truth; and one might add that Gödel was personally a Platonist, worlds removed from postmodernist skepticism.

My second example has to do with quantum theory, which can be viewed as entailing a limit theorem of a very different kind. What first comes to mind is the Heisenberg Uncertainty Principle, which limits the accuracy with which the values of conjugate dynamic vari-ables of a quantum system (such as position and momentum) can be ascertained. What stands at issue, as one believes today, is not simply an incapacity on the part of the experimentalist, but the fact that dynamic variables of a quantum system do not, in general, *have* a definite value. An electron, for example, may not have a definite position or a specific momentum, and in any case, can never have both at once. It follows that Heisenberg Uncertainty restricts the applicability of pre-quantum physics to a macroscopic domain within which quantum effects can be neglected. And this is one way in which quantum theory can be seen as entailing a limit theorem.

On closer examination, one finds however that more is at stake; as I have shown at length in the first two chapters, quantum mechanics entails an ontological distinction between the physical and the corporeal domains. To be precise, it is the phenomenon of state vector collapse that demands what may be termed the rediscovery of the corporeal world, a domain which in principle falls beyond the reach of physical science as such. What confronts us here is a limit theorem so radical and so profound, that physicists have as yet been unable to come to grips with its implications.

Yet there is more to be said: as I have shown in Chapter 3, quantum mechanics entails in addition a limit theorem which affirms the boundedness, not only of the physical domain, but of the spacetime continuum that encompasses the corporeal world itself. In light of traditional ontology, this is tantamount to a rediscovery of the so-called intermediary domain. I cannot but agree with Henry Stapp that this constitutes indeed "the most profound discovery of science": for it takes us, ontologically speaking, to the deepest stratum of cosmic reality the existence of which science can detect.

Turning now to other sciences, we encounter next a very different kind of limit theorem in William Dembski's theory of "intelligent design" with which we have dealt at length in Chapter 9. What the pivotal theorem proves is that neither chance, nor necessity, nor the combination of the two, covers the entire ground of causality. It demonstrates, in particular, that what we normally recognize as instances of "design" cannot in fact be ascribed to the aforesaid modes of causation. Now these findings, clearly, constitute a causal limit theorem which restricts the domain of phenomena that can be accounted for in scientific terms. It happens, moreover, that the entire biosphere constitutes a case in point, which is to say that in general the origin of biological structures cannot be ascribed to natural causation. Thus if quantum theory has brought to light the existence of higher ontological planes, Dembski's theorem demonstrates that neither the physical universe nor the corporeal (nor the two combined) constitute what scientists term a closed system.

My next example concerns the theory of visual perception proposed by the late James J. Gibson, a Cornell University psychology professor who devoted fifty years of his life to the study of how we perceive.[2] By way of painstaking empirical investigations he became convinced that the prevailing theories of visual perception prove to be in fact untenable. The very notion that "the eye sends, the nerve transmits, and a mind or spirit receives" needs to be radically modified. In the final count, perception is to be conceived as an act, not of the body, nor of a mind, nor indeed of the two operating in tandem, but of the mind-body compound, conceived holistically as a single entity. What Gibson terms the perceptual system is not a sum of parts, nor can the perceptual act be dichotomized into stimulus and response. And as to the famous "perceptual image"—whether conceived as existing physiologically in the brain or psychologically in the mind—he concludes that the concept is spurious.[3] What is perceived, Gibson finds, is not an image, but quite simply the external environment; in a word, the so-called *ecological* theory of perception is non-bifurcationist. "This distinction between primary and secondary qualities is quite unnecessary," writes Gibson, and is in fact "wholly rejected" in his approach.[4] One is amazed to see how this sober scientist was able, by way of hard-headed inquiry based squarely upon empirical findings, to deconstruct the Cartesian edifice. He shows that the customary neurological and computer-theoretic approach to perception is flawed, and can at best yield results

2. See *The Ecological Approach to Visual Perception* (Hillsdale, NJ: Lawrence Erlbaum, 1986). For a summary and analysis of Gibson's theory I refer to chapter 4 in *Science and Myth*, op. cit.

3. Ibid., pp. 60–61.

4. Ibid., p. 31.

of a secondary nature—a recognition which can be seen as a decisive limit theorem pertaining to cognitive psychology.

There is however more to be said; for it happens that visual perception is but a special case of something far more general: a "seeing" namely, which constitutes the quintessential act in all forms of human cognition. To know an object of whatever kind is finally to *see* it in intellective mode: the intellect, as the ultimate agent of human knowing, proves thus to be indeed "the eye of the soul."

The question arises now whether that intellective seeing, too, may be subject to a limit theorem: if visual perception proves intractable to the methods of contemporary science, is it not reasonable to suppose that the same holds true when it comes to that incomparably more basic "seeing" which consummates all knowing *per se*? The problem, of course, is to find a way to attack this question by scientific means, and what finally renders the subject accessible to rigorous scientific inquiry is the fact that it includes *mathematical* knowing—the recognition, namely, of mathematical truth—as a special case. For when it comes to the validation of mathematical theorems one has at one's disposal a science of the most rigorous and formidable kind. It should not be altogether surprising, therefore, that the very considerations of a meta-mathematical kind which have yielded our first limit theorem will also yield our last.

It was Roger Penrose, the Oxford mathematician and mentor of Stephen Hawking, best known perhaps for the Hawking-Penrose Singularity Theorem, who made the discovery. Following his exploits in astrophysics he turned to the study of the human brain, a field presently dominated—for very good reason—by the computer paradigm. Yet Penrose began soon to entertain doubts regarding the scope of that paradigm—and where better to "check this out" than in the case of *mathematical* knowing! To this end he recalled the crux of Gödel's argument, which consists in constructing, by exceedingly artful means, a *true* arithmetical proposition P which however admits no formal or "algorithmic" proof. At the end of an intricate argument entailing a lexicographical ordering of *all* arithmetical propositions one arrives at a particular proposition P which is both *unprovable* and *true*. Now, as Penrose recognized instantly, the implications of this logical fact are immense: for it proves—once and for all!—that mathematical knowing is *not* algorithmic, not something that could in principle be accomplished by a computer, be it in the form of a brain. What is called for, Penrose concludes, is indeed a "seeing" effected by a non-algorithmic act: "We must 'see' the truth of a mathematical argument" he tells us. But there is still more to be said: for if such a non-algorithmic act proves to be essential in the discernment of mathematical truth, how much more must this be the case when it comes to non-mathematical judgments! Pondering the wider significance of his Gödelian discovery—his limit theorem—Penrose concludes that "this 'seeing' is the very essence of consciousness."[5] It is to be noted that the Oxford mathematician has in a way generalized Gibson's limit theorem from visual perception to human knowing *per se*: in effect he has rediscovered what is properly termed *intellect*.

Hard science, it turns out, is ultimately destructive of scientistic myth. The scientific enterprise is inherently self-corrective: hard science, as I have said, is wiser, in a sense, than the individual scientist. Consider the phenomenon of state vector collapse, for instance: this finding came not only as a complete surprise, but as a shock to the scientific community. Erwin Schrödinger, one of the founders of quantum theory, was so disturbed by this phenomenon—"this damned jumping" as he called it—that he rued his own discovery. One sees, in particular, that the limit theorems of quantum mechanics—or better said, the findings which underlie these theorems—have thrust themselves upon the scientific community by force of an inexorable logic, which not even the might of an Albert Einstein could thwart.

We need also, however, to note the following: If hard science is indeed wiser in a sense than even the greatest scientists, it is wiser too than the postmodernist philosophers who impugn the enter-

---

5. *The Emperor's New Mind* (Oxford University Press, 1990), p. 418. The argument itself may be found on pp. 99–112.

prise. I am not denying for a moment that the new philosophy of science has contributed major insights: scientific facts, for example, may indeed be "theory laden" as I have noted before. What strikes me as objectionable in the postmodernist outlook, on the other hand, is its pervasive relativism, which attacks truth the way an acid eats up metal. To be sure, this is hardly the place to articulate a critique of postmodernism; suffice it to say that Frithjof Schuon may well have hit the nail on the head when he observed that "its initial absurdity lies in the implicit claim to be unique in escaping, as if by enchantment, from a relativity that is declared alone to be possible." Meanwhile science is continuing to evolve, continuing to unfold its possibilities, and has now attained levels of discovery that shake the foundations of scientistic belief. My point is that hard science has by now disavowed the very premises which gave rise, first to modernist enlightenment, and two centuries later, by way of reaction, to postmodernist skepticism.

What then? If the postmodern, as well as the modern outlook, have now been discredited, what viable option remains? The answer can be given most succinctly in the words of Alexander Solzhenitsyn when he declared before a Harvard audience (to the bewilderment of students and faculty alike): "Today the only way left is *upwards.*"

But this way, surely, cannot be improvised, drawn out of thin air: it needs rather to be rediscovered, appropriated, *received.* The very concept of "verticality" is foreign to modernity and postmodernity alike; it reaches us from a past we have been taught to despise as primitive and superstitious. We have forgotten, most of us, that *tradition* can be more than a custom, a convention, a mere remnant to be discarded at will. That such "vestiges" can be enlightening, that they may serve to deepen our understanding—this we find hard to believe. We do not think highly of human culture in this Darwinist age. My point is simple: It is high time to pay serious and respectful heed once more to the venerable pre-modern traditions; instead of turning postmodernist, let us open our minds to the transmitted wisdom of mankind.

What is called for above all is a rediscovery of traditional cosmology. Culture and cosmology, as I have said before, go hand in hand; and what is lacking in the modern West—in both our culture and our cosmology—is precisely the dimension of verticality. As our universe flattens, so does our conception of man, and so does our culture in all its aspects.[6] There are compensations, of course, on various horizontal planes; but these do not suffice: man was not born for that: he absolutely needs the vertical dimension to be fully human. There are those, to be sure, who accept this latter tenet, but would question that it has anything to do with cosmology. Verticality refers to an inward dimension, they would contend; it refers to something spiritual with which cosmology has no concern. But the matter is not quite so simple—nor quite so Cartesian, in fact. The inner and the external, it turns out, are profoundly linked. I repeat: As our universe flattens, so does our culture; and as Huston Smith points out: "A meaningful life is not finally possible in a meaningless world."

We stand in need of a new cosmology: of a cosmos incomparably more vast than the universe of contemporary astrophysics. I am not of course referring to spatial extension: the physical universe as currently conceived encompasses light-years enough! I speak rather of things which cannot be measured or weighed, of things, in fact, which can only be spoken of in traditional and symbolic terms: of an integral cosmos made up of distinct ontologic levels, which we may picture as so many horizontal planes or concentric spheres.[7] I speak thus of a cosmic hierarchy, a universe with an added dimension: the dimension of verticality, which has to do, not with spatial direction, but with value and meaning, and ultimately, with first origins and last ends. It is the dimension that transforms the cosmos from a mere *thing* into a *bona fide* symbol: a theophany, no less; it is thus the dimension that nourishes the artist, the poet, and above all, the mystic in us—the dimension, as I have said, which enables us to be fully human. It is also, however, the dimension that in fact permits existence, permits *being* as such; for it can indeed be said, onto-

---

6. I have elucidated this contention in *Cosmos and Transcendence*, op. cit., Chapter 7.

7. An excellent account of hierarchic cosmology has been given by Seyyed Hossein Nasr in his 1981 Gifford Lectures. See *Knowledge and the Sacred* (New York: Crossroad, 1981), which also gives extensive references.

logically speaking, that nothing can exist simply on a horizontal plane. In the absence of verticality, therefore, nothing can be understood, nothing can be known in truth. It is no small disadvantage, thus, that verticality has been banished in modern times: postmodernist nihilism, it turns out, is by no means unjustified. In fact, it is profound. Nietzsche was right: "We have abolished the true world. What has remained? The apparent one perhaps? Oh, no! With the true world we have also abolished the apparent one"—prophetic words!

But the question remains whether the true world can be reinstated. To be precise: Can hard science sanction a multi-level cosmology of the traditional kind? I submit that it can. What is more, I contend that science today not only permits a hierarchic cosmology, but in fact *demands* a worldview of this kind; it is only that science as such cannot *articulate* that demand. As we have seen in light of the perennial ontology, quantum mechanics itself entails three distinct ontological levels: the physical, the corporeal, and the intermediary. Turning to Dembski's theory, one finds that the newly-discovered criterion of design permits us to distinguish scientifically between the animate and inanimate levels of corporeal being, in accordance with traditional cosmology. To this hierarchic structure now comprising *four* tiers, Gibson's "ecological" theory of perception entails a further division through its discernment of perception as an act *sui generis*, irreducible to physiology. It thus appears that in its own way the theory distinguishes between the plant and animal levels within the biosphere. I would argue further that the Penrose theorem—which as I have suggested before is tantamount to a rediscovery of what is properly termed *intellect*—distinguishes the human domain categorically within the realm of animals. In any case, one thus recovers four links of what Arthur Lovejoy termed "the great chain of being," with a fifth, comprised of the physical domain, added to the lower end.

To be sure, our scientific grip loosens the higher we ascend along this chain: it must, seeing that we are ascending towards the pole of *essence*, the very thing our Baconian sciences are by nature unable to grasp. The wonder is that they apparently comprehend enough of supra-physical reality to validate the aforesaid limit theorems. In accomplishing this task, however, they have reached the end of their tether. Geared as they are to the quantitative aspects of reality, which are rooted in the material pole of existence, their actual purview does not extend beyond the corporeal domain. We should point out in this connection that the ontological discoveries enumerated above derive invariably from *negative* results: from limit theorems which disclose impenetrable boundaries for the science in question. It needs however to be realized that these scientific findings reveal their *ontological* implications only by way of traditional doctrine. To be precise, it is in light of this metaphysical knowing that the limit theorems of contemporary science reveal an *ontological* distinction between the physical, corporeal and intermediary degrees, as well as a subdivision of the corporeal corresponding to the traditional "mineral, plant, animal" trichotomy. The point, however, is that these ontological recognitions transcend what the scientist as such can understand. To proceed beyond limit theorems as such, one requires means of a very different kind. Above the physical domain, *essence* comes into play; and this is what a science based upon the discernment of quantity is categorically unable to comprehend. *The only way to grasp essence is through an act of perception, be it sensory or intellective.* What the descendents of Descartes have roundly forgotten is that by virtue of essence the human subject does have access to the external world, and that the miracle is in fact accomplished by every man, woman or child endowed with "eyes to see." It is moreover to be understood that such seeing is basic to the *modus operandi* of the traditional sciences, which are ultimately concerned, not simply with quantitative parameters, but with *essences*, precisely.

One might remark that the conception of a science based upon "seeing" has not altogether disappeared from the contemporary West. Such a discipline was in fact championed by Johann Wolfgang von Goethe during the heyday of the Enlightenment; and while that Goethean science was derided and ultimately ignored for more than a century, it has of late been rediscovered and is presently being applied in various domains. I have elsewhere mentioned the name of Henry Bortoft, the physicist whose book *The Wholeness of Nature*—subtitled "Goethe's Way toward a Science of Conscious

Participation in Nature"—constitutes a remarkable tribute to the depth and range of that long-neglected discipline. It should moreover be noted that "conscious participation in Nature" is evidently to be achieved by means of direct perception, which as we have said before, constitutes the one and only basic means of grasping a cosmic entity of whatever kind in its essence.

One sees finally that contemporary science, stringently limited though it be in what it is able to know, endorses nonetheless a hierarchic worldview of the traditional kind. A restoration of cosmology, unthinkable a century ago, has thus become theoretically feasible. Since the Enlightenment, Western man has found himself intellectually in a flattened cosmos, a truncated universe of mere particles, persuaded that science had so decreed; and now one knows—or ought to know!—that we have been deceived. It was scientism, it turns out, that perpetrated the fraud; and this we now know on the authority of science itself. Strange as it may seem, science as such, by means of its limit theorems, has cleared the way for the rediscovery of a timeless wisdom: a bona-fide knowledge concerning the cosmos at large, which transcends what science itself can grasp. The way has now been cleared for the restitution of a *vertically*-oriented mode of knowing, answering once again to the deeper aspirations of mankind.

Finally, let it be said that if this fact were recognized—if it were comprehended by our pundits and affirmed in our universities—the modern world as such would forthwith cease to exist, and a restoration of culture would then be feasible.

INDEX OF NAMES

Made in the USA
Monee, IL
17 May 2023

33913503R00152